BATTLESHIPS OF WORLD WAR I & WORLD WAR II
1914–45

BATTLESHIPS OF WORLD WAR I & WORLD WAR II
1914–45

E.V. MARTINDALE

First published in 2023

Copyright © 2023 Amber Books Ltd

All rights reserved. No part of this publication may be reproduced, stored in a retrieval system, or transmitted in any form or by any means, electronic, mechanical, photocopying, recording, or otherwise, without prior written permission of the copyright holder.

Published by Amber Books Ltd
United House
London N7 9DP
United Kingdom
www.amberbooks.co.uk
Facebook: amberbooks
Instagram: amberbooksltd
Twitter: @amberbooks
Pinterest: amberbooksltd

ISBN: 978-1-83886-294-7

Editor: Michael Spilling
Designer: Andrew Easton
Picture research: Terry Forshaw

Printed in China

Contents

Introduction 6

World War I 8

World War II 78

GLOSSARY 122
INDEX 124
PICTURE CREDITS 128

Introduction

To achieve its purpose in combat, a twentieth century battleship's three essential elements were its armament, its armour, and its speed. With very few exceptions, all these ships were powered by steam.

Earlier in the century they were coal-fired, with triple-expansion reciprocating engines. By 1915, the most advanced ships were oil-fired and driven by geared steam turbines. Throughout the period, designers sought the harmonization of conflicting demands. Speed needed multiple boilers to raise steam, fuel bunkers to feed them, and very large and powerful engines, all of which took up a lot of space. Large-calibre guns were very heavy, and had to be positioned to have an effective field of fire and to maintain the ship's balance. Solid armour along the sides was necessary to protect the machinery and ammunition magazines against increasingly effective shell-fire and torpedoes. More weight in turn required more power to push the vessel through the water. Unsurprisingly, the ships grew in size as each navy sought to gain the advantage. In the 1900s, a capital ship might have a displacement of 14,000 tonnes; by 1910, this had reached 20,000 tonnes, and the battleships of 1916–17 exceeded 30,000 tonnes. Britain's *Hood* (1920) topped 40,000 tonnes, Japan's rebuilt *Fusō* (1933) matched this, the United States' *Missouri* (1944) displaced 53,000 tonnes fully loaded; and Japan's *Yamato* (1940) was the biggest ever at 72,806 tonnes fully loaded.

The German battlecruisers *Seydlitz* and *Derfflinger* manoeuvre accompanied by torpedo boats during the Battle of Jutland, 1 June 1916.

INTRODUCTION

Power projection

These great increases in the dimensions of capital ships reflected both the strategic conservatism and the technical innovativeness of the main naval powers. Battleships got bigger so that they could wield greater destructive power, because they required more space for advanced technology in gunnery and communications, because they could then carry more fuel and ammunition and so have greater endurance at sea; also because they came to need strong anti-aircraft defence. However, with the possible exception of Nazi Germany, there was little discussion of what these ships were actually for.

The naval arms limitation treaties of 1922 and 1930 may have played a part here. Their emphasis on limiting the size of the largest ships helped to confirm the status of the battleship as the prime unit of a navy, though many observers noted in the 1930s that the submarine and the aircraft carrier would radically alter the dynamics of naval warfare in the next war.

But as long as one navy had an operational battleship, rival or enemy navies felt the need to maintain their own. In the later stages of World War II battleships – apart from the US Navy in the Pacific – resembled kings on a

A Royal Aircraft Factory S.E.5 airplane sits on top of number three turret of USS *Mississippi* (BB-41) while testing basing aircraft on battleships, 3 June 1919.

chessboard, vital symbols of power to be conserved and protected while lesser pieces fought. Navies continued to build them right up to the end of the war, but at an increasingly slow pace. The urgent need for aircraft carriers and convoy escorts took priority. The last battleship to be commissioned, the Royal Navy's *Vanguard* (1946), was used largely for ceremonial purposes until it was placed in reserve in 1955 and finally scrapped in 1960.

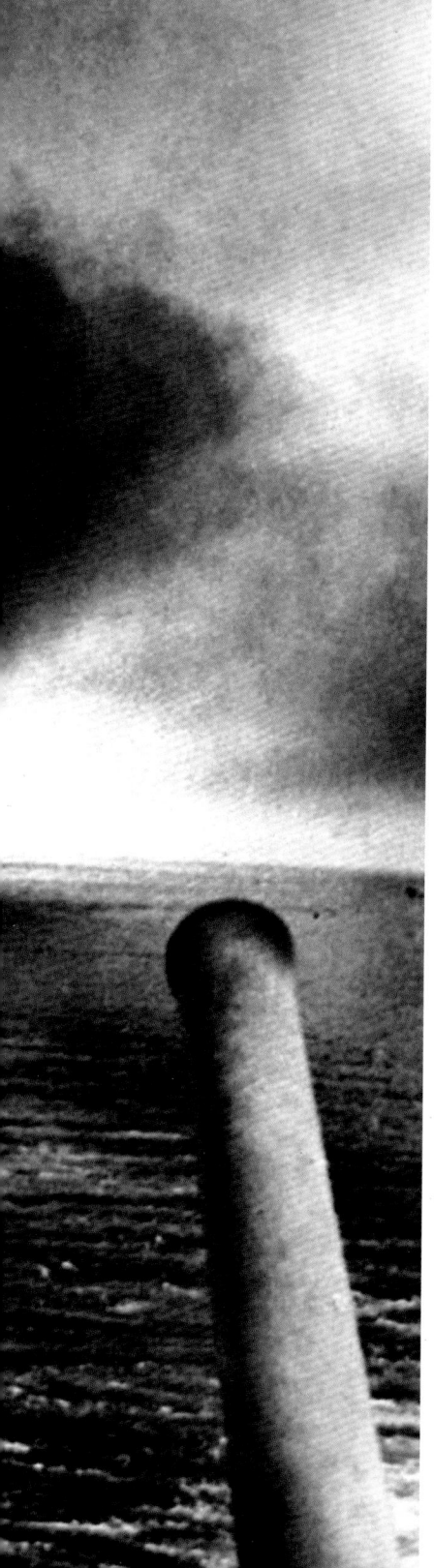

WORLD WAR I

The oldest battleship to participate actively in World War I was the French Navy's *Jauréguiberry*, which was first commissioned in 1896 and acted as flagship of the French squadron at Gallipoli in 1915. A few others, such as the USS *Iowa*, the German *Kaiser Friedrich III* and the Royal Navy's *Canopus*, saw some active service, but all pre-1900 vessels were withdrawn before the end of the war, to be used for accommodation, training or storage, or they were simply scrapped. The war at sea was largely conducted by ships built in the rush to create new battle-fleets that began in 1906. These were larger, faster and more heavily armoured and gunned than their predecessors. Yet, as the demands of warfare pushed forward new developments in weaponry, fresh vulnerabilities were also becoming clear, including the human difficulty of tactically managing formations of such large ships in action across miles of open sea.

Lessons of Jutland

These ships were very different to the 'first rates' of the early nineteenth century. Villeneuve's flagship at Trafalgar, *Bucentaure*, displaced 1455 tonnes (1432 tons), Nelson's *Victory* (40 years old in 1805) displaced 1961 tonnes (1930 tons). Yet the war's main sea battle at Jutland on 31 May–1 June 1916 was, while faster-moving and more widely spaced, essentially conducted on the same tactical basis: of two fleets approaching each other in line of battle and endeavouring to break through the enemy's line and encircle its ships with concentrated gunfire. The speed and manoeuvrability of twentieth-century warships made this an impossible task and resulted in the inability of either side to claim a clear victory. The lessons of Jutland would not be lost on other navies, not least that of Japan.

Important advances in technology were made in the 1900s. Effective ship management depended increasingly on mechanical and electrical engineering, and this was reflected by the rather belated decision of the British Admiralty to create a proper hierarchy of engineer officers with equivalent status to the navigational cadre. By 1903, the Navy List specified an officer's branch: N (navigation), T (torpedo), G (guns), E (engines). Naval ships in 1899 had coal-fired, fire-tube boilers supplying steam to reciprocating engines. None had radio. Ten years later, new ships, although still largely coal-fired, had water-tube boilers that powered steam turbines, and all major ships had radio. Fire control had improved, and the need to house control stations was changing the profile of battleships. However, the very concept of the battleship was challenged, or augmented, by the introduction of the battlecruiser.

The German battleship *Schleswig-Holstein* firing a salvo during the Battle of Jutland (North Sea, 31 May–1 June 1916).

WORLD WAR I

Canopus (1899)

UNITED KINGDOM

HMS *Canopus* was the lead ship of a class of six, each one constructed in a different dockyard. Although planned for deployment in the Far East, only one (*Goliath*) ever got as far as the East Indies Station.

Laid down in January 1897 and completed in December 1899, *Canopus* – like its sisters – was fated for early obsolescence. The class was a more modern version of the almost-contemporary HMS *Majestic*, with lighter armour and more powerful engines. The design was led by the Director of Naval Construction, Sir William White. The maximum laden draught of 9.1m (29ft 10in) was intended to allow passage through the Suez Canal.

Boiler controversy

The class was at the centre of a controversy over the Admiralty's decision to switch from cylindrical fire-tube boilers to water-tube boilers, in which the water flows through tubes inside flues that contain extremely hot gases. This reverses the arrangement of the previously used 'Scotch' marine boiler, in which the gases passed through fire-tubes, with the water surrounding them. The concept was not new but its development intensified with the introduction of the steam turbine, which required steam at both greater heat and pressure. *Canopus*, however, like all pre-*dreadnought* battleships, had reciprocating engines, which were less demanding on coal: an important consideration for endurance time at sea.

Early water-tube boilers on British warships had a poor record of efficiency, but the modernist element at the Admiralty put this down (correctly) to a combination of defective manufacture and unsatisfactory management of the boilers on board ship. A further consideration was that water-tube boilers needed more intensive firing, which meant that deck crew had to assist with stoking. This, of course, was undesirable if the ship should be in action. *Canopus*, with a bunker capacity of 2032 tonnes (2000 tons), with 20

Canopus
Dimensions: Length 128m (420ft), Beam 23m (75ft), Draught 9.1m (29ft 10in), Displacement 13,360 tonnes (13,150 tons)
Propulsion: 2 vertical 3-cylinder triple expansion engines, twin screws, 10,067kW (13,500shp)
Armament: 4 305mm (12in) guns, 12 152mm (6in) guns, 10 12-pounder guns, 4 457mm (18in) torpedo tubes
Armour: Belt 152mm (6in), Barbettes 305mm (12in), Deck 51mm (2in)
Range: 14,820km (8000nm)
Speed: 18 knots (33.33km/h)
Complement: 694

'Stern walk'
The railed 'stern walk', a legacy from wooden-hulled warships, lasted up to the later 1890s.

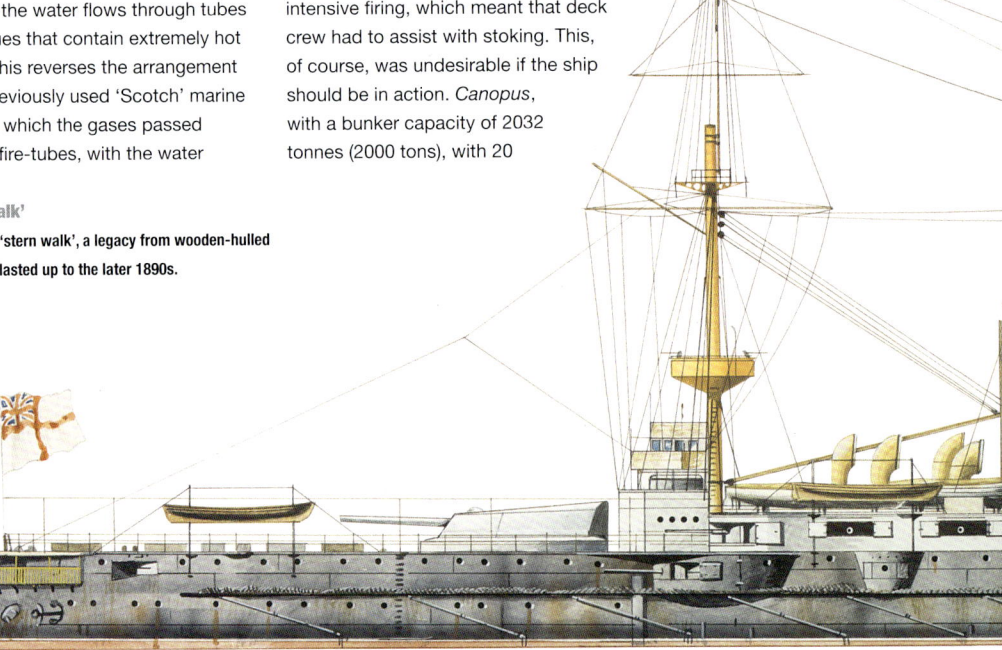

Belleville boilers, burned over 10 tonnes (11 tons) an hour at full speed.

As with *Majestic*, the main armament was four 305mm (12in) guns mounted in twin turrets, fore and aft, and twelve 152mm (6in) guns in single semi-revolving turrets mounted in side casemates – an almost universal arrangement of the time, although it meant that the secondary batteries were often awash in a choppy sea. The fore and aft 152mm turrets were mounted on sponsons extending from the hull, to enable a wider range of fire.

Canopus was among the first RN ships to carry radio apparatus, fitted in 1900 with Marconi gear that enabled it to exchange transmissions with another ship up to 48km (30mi) distant.

Canopus c. 1899. 'Battleship grey' was adopted for hulls and upper works only in 1903.

Victory at Port Stanley

Canopus's big moment came as guardship at Port Stanley in the Falkland/Malvinas Islands, where it was sent in October 1914. On 8 December Admiral Maximilian, Graf von Spee's German squadron was approaching the islands, and *Canopus*, temporarily beached in an inlet, fired its main guns over an intervening hill, causing Spee to head out seawards again. A British battlecruiser force sent to find and destroy Spee's ships was coaling at Port Stanley, unknown to the Germans. With superior armament and speed, they pursued and sank the German ships. *Canopus* returned to England in January 1915 and was again in action in the Dardanelles campaign that same year.

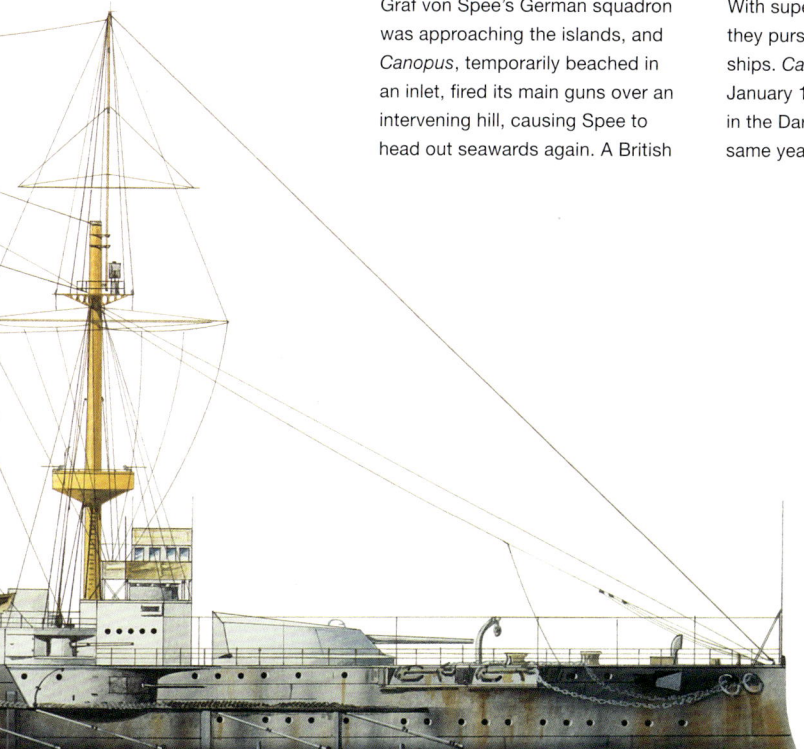

WORLD WAR I

Kearsarge (1900)

UNITED STATES

Classified as BB-5, *Kearsarge* with its sister ship *Kentucky* was intended for coastal defence. Laid down at Newport News on 30 June 1896, launched on 24 March 1898 and commissioned on 20 February 1900, it had a 55-year career, although only 20 of those years was as a battleship.

The armoured conning tower was surmounted by a glazed wooden navigation bridge with open railed wings, and the original pole-type masts, each with two spotting/light gun platforms were replaced by basket-type masts during a refit in 1909.

With five boilers, albeit of the double-ended Scotch type, the ship was somewhat under-powered. Its triple expansion engines, with a power output of 7457kW (10,000shp), gave it a modest top speed of 16 knots (see USS *South Carolina*). Maximum bunker capacity was 1617 tonnes (1591 tons), although this was only for long cruises.

Gun power

As designed, the ship had a higher freeboard than the preceding *Indiana* class (4.42m/14ft 6in), but even so the 127mm (5in) side batteries were almost unusable in anything more than a moderate sea. The main guns, while quick-firing, had to be set to an elevation of 2° for reloading. The turret design, by an officer in the Bureau of Ordnance, Joseph Strauss (later Admiral), placed the 203mm (8in) guns directly on top of the 330mm (13in) guns, giving them no independent training. Although also used in the subsequent *Virginia* class, this arrangement was not successful, however, it was a stepping stone towards the superfiring turret. Strauss was also responsible for developing the electric operation of heavy gun mounts, which became standard in the US Navy. There was much discussion about the most effective gun calibre at the time, but *Kearsarge* was the only US ship to carry 330mm (13in) guns as main armament, and 305mm (12in) remained the big-gun standard until 1914.

US battleships of this era carried a variety of boats: in *Kearsarge*'s case, 6 cutters, 2 launches, 1 admiral's barge, 2

Deck levels
The low sheer typical of US battleships of the period is evident here, with the weather deck at the same level from stem to stern.

WORLD WAR I

whaleboats, 1 gig, 2 dinghies and 2 catamarans.

With short fore- and after-decks, it was tubby in plan, rapidly widening to 22.02m (72ft 3in), which made it very suitable for conversion to a crane barge, classed as AB-1 in August 1920. With all superstructure removed, the hull carried a revolving crane with a lifting capacity of 254 tonnes (250 tons), and in this form it worked in Navy yards until 1955. The name was transferred to a carrier in 1941, and the ex-battleship became Crane Ship No. 1.

White paint was used for the US Navy's "Great White Fleet" world tour of 1907, but the US Navy reverted to grey by 1910.

USS Kearsarge
Dimensions: Length 131.7m (432ft), Beam 22.02m (72ft 3in), Draught 8.28m (27ft 6in), Displacement 15,444 tonnes (15,200 tons)
Propulsion: 25 Belleville water-tube boilers, 2 vertical triple-expansion engines, two screws, 7457kW (10,000shp)
Armament: 4 x 330mm (13in) guns, 4 x 8 in (203mm), 14 x 127mm (5in) guns, 20 x 12-pounder, 8 x 3-pounder, 4 x 2.5-pounder guns, 4 x 457mm (18in) torpedo tubes
Armour: Belt 229–102mm (9–4in), Bulkheads 305mm (12in), Deck 76mm (3in), Barbettes 356–254mm (14–10in), Main turrets 254–203mm (10–8in), Lower deck redoubt and battery 152mm (6in)
Range: 13,000km (8125mi, 7060nm) at 10 knots (18.5km/h/11.5mph)
Speed: 18 knots (33.3km/h, 20.7mph)
Complement: 830

WORLD WAR I

Mikasa (1902)

JAPAN

Now a museum ship, *Mikasa* is the only surviving pre-*dreadnought* battleship in the world. Japan's 1896–97 naval plan centred on the building of four battleships, all in British yards.

Mikasa was the fourth and in many ways an improved version of the British *Majestic* class, completed by the Vickers yard at Barrow-in-Furness in 1902. It followed the convention of the time with a ram-type stem, secondary guns in lateral casemates and a mixed array of guns. Military-style pole masts carried the lightest 2.5 pounder guns. In these and other ways the ship was designed for relatively close combat. Yet it was its most modern features, including fire control and radio, that were vital to the Japanese victory in the fast-moving, relatively long-range Battle of Tsushima on 27–28 May 1905, when *Mikasa* was Admiral Togo's flagship.

Adopting KCA
The hull was fitted with Krupp cemented armour (KCA), developed in Germany. The steel plates had a hardened outer face combined in the manufacturing process with a more fibrous and elastic inner surface. Alloys in the steel were carbon (0.35 per cent), nickel (3.9 per cent), chromium 2 per cent, manganese 0.35 per cent, silicon 0.07 per cent, phosphorus 0.025 per cent and sulphur 0.02 per cent. KCA offered the same protection as Harveyized armour, but with 20 per cent less thickness and a consequent reduction in armour weight.

Planning for Tsushima
With a new and growing fleet of battleships, the Imperial Navy

Ventilation tubes
Cowled ventilation tubes, to capture the wind, were still in use at this time. Electric fans came into use around 1910.

devoted intense thought to battle planning, with a major contribution from Akiyama Saneyuki, a staff officer who had studied US naval strategy and tactics. The main battery guns could be worked hydraulically, electrically or even manually, could be loaded at any elevation, and had a rate of fire of three shells every two minutes. From 1903, all Japanese naval ships larger than destroyers had radio with a 112km

(70mi) range. Although *Mikasa* took some 40 strikes, including 10 from 305mm (12in) shells, it suffered no major damage.

Tsushima was the first battle in which radio communication between ships was important. It also demonstrated the value of heavy guns, long-range fire (with rangefinders to guide the guns) and speed. A battleship's secondary armament of 203mm (8in) or 152mm (6in) guns was shown to be ineffective and perhaps even detrimental to successful combat. It was the 305mm (12in) guns, firing at relatively long range that decided the issue.

Mikasa

Dimensions: Length 131.7m (432ft), Beam 23.2m (72ft 6in), Draught 8.28m (27ft 6in), Displacement 15,444 tonnes (15,200 tons)

Propulsion: 25 Belleville water-tube boilers, 2 vertical triple-expansion engines, two screws, 11,185kW (15,000shp)

Armament: 4 305mm (12in) guns, 14 152mm (6in) guns, 20 12-pounder, 8 3-pounder, 4 2.5-pounder guns, 4 457mm (18in) torpedo tubes

Armour: Belt 229–102mm (9–4in), Bulkheads 305mm (12in), Deck 76mm (3in), Barbettes 356–254mm (14–10in), Main turrets 254–203mm (10–8in), Lower deck redoubt and battery 152mm (6in)

Range: 13,000km (8125mi, 7060nm) at 10 knots (18.5km/h/11.5mph)

Speed: 18 knots (33.3km/h, 20.7mph)

Complement: 830

Open bridge deck

The full-width open bridge deck was a common feature among navies, enabling officers to get a clear view fore and aft on each side. It lasted longest on US and German battleships.

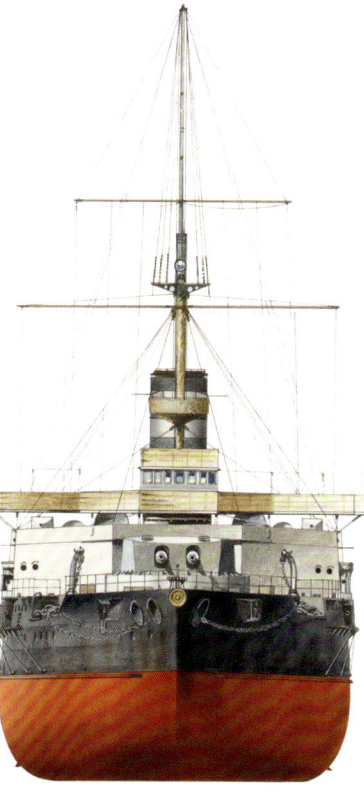

WORLD WAR I

Tsesarevich (1903)

RUSSIA

The imperial Russian naval authorities considered the French to be the most advanced battleship designers. *Tsesarevich* **was not only a French design but it was built in a French yard at La Seyne-sur-Mer, and commissioned into the Russian Navy on 3 September 1903.**

Its appearance somewhat resembles that of a modern 'stealth' ship. However, its almost total enclosure was a protection against shells, not to minimize its radar image for radar was still three decades away. With a narrow deck area, sponsons were fitted to the sloping hull to give the 152mm (6in) guns a wider field of fire.

The inwards-sloping 'tumblehome' design was something of a fetish among French naval designers of this period, although only the Russians also adopted it. Intended to improve stability, it could cause the ships to roll and wallow in a swell. By the advent of the *République* class in 1906, it was discarded. A long internal watertight bulkhead separated the two engine rooms, and helped save the ship when it was struck by a Japanese torpedo in the attack on Port Arthur (8 February 1904).

Armoured deck
Forecastle and afterdeck were notably short. Hull and superstructure appeared to merge into each other, but there was a full-length armoured deck 57mm (2.25in) thick, laid on the normal deck plating. A series of cells, mostly used as coal bunkers, were fitted inside the hull to add to the external armour. Twenty Belleville boilers at a working pressure of 17.6kg/cm^2 (250psi) gave steam to two vertical

Torpedo tubes
The caps of the bow and stern-mounted torpedo tubes can be seen just below the waterline. The stem shape implies an underwater ram but none is fitted.

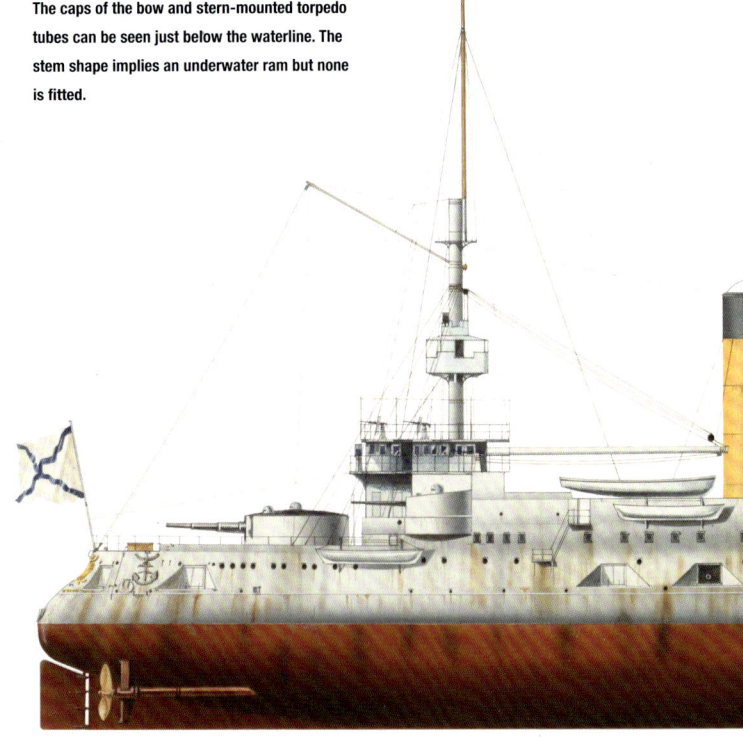

triple-expansion engines and also to six electric generators that powered the gun turrets, with a total capacity of 550kW (740shp).

Model battleship

Tsesarevich became the model for the slightly larger Russian-built battleships of the Borodino class. It was active in the Russo-Japanese war of 1894–95 and was hit by a torpedo on the port quarter on 8–9 February 1894. Its protection limited the damage and deliberate flooding of compartments on the starboard side kept it stable. It was severely damaged again in the Battle of the Yellow Sea (10 August 1894). Taking refuge in the German-controlled port of Tsingtao, it was interned but later returned to Russia.

Longest-serving battleship

It was the only Russian battleship to serve both in the Russo-Japanese War and in World War I, when it served in the Baltic Fleet. On 13 April 1917, after the Russian Revolution, it was re-named *Grazhdanin* ('Citizen') and took part in the battle against German ships at Moon Sound (October–November 1917). *Tsesarevich* was hulked at the Kronstadt naval base in May 1918 and scrapped in 1924.

Deck plan
Tsesarevitch was a compact and effective fighting machine but its tubby shape caused it to heel over dangerously when making turns.

Tsesarevich

Dimensions: Length 118.5m (388ft 9in), Beam 23.2m (76ft 1in), Draught 8.5m (27ft 11in), Displacement 13,122 tonnes (12,915 tons)
Propulsion: 20 Belleville boilers, 2 vertical triple-expansion engines, 12,155kW (16,300shp), 2 screws
Armament: 4 305mm (12in) guns, 12 152mm (6in) guns, 16 75mm (3in) guns, 4 47mm (1.9in) 3-pounder guns, 6 457mm (18in) torpedo tubes
Armour: Belt 230–150mm (9–5in), Main turrets 250mm (10in), Secondary turrets 150mm (5.9in), Conning tower 250mm (10in), Deck 57mm (2.25in)
Range: 10,186km (5500nm) at 10 knots (18.5km/h/11.5mph)
Speed: 18 knots (33.3km/h, 20.7mph)
Complement: 779

WORLD WAR I

Dreadnought (1906)

UNITED KINGDOM

The speed of *Dreadnought's* construction and fitting-out by the Portsmouth Navy Yard – 14 months from laying the keel to commissioning on 2 December 1906 – was hardly less impressive than the ship itself.

Its planning and construction actually coincided with a drive to reduce the cost of new warships, and part of the design brief was to provide the maximum gunpower possible on a hull of around 17,000 tonnes (16,731 tons). Its cost of £1,785,683 compared favourably with the preceding *Lord Nelson* class (£1,651,339). The key to *Dreadnought's* success was that it integrated the strands of current thinking about a battleship's role into a single ship.

Novel design

The main novelties of the design consisted firstly, of 10 big guns all of the same calibre and no intermediate armament; and second turbine propulsion that gave it increased

---

Dreadnought

Dimensions: Length 160.4m (527ft), Beam 25m (82ft), Draught 8.1m (26ft 6in), Displacement 18,186 tonnes (17,900 tons); 22,195 tonnes (21,845 tons) full load

Propulsion: 18 Babcock & Wilcox boilers, 4 direct-drive Parsons turbines, 17,000kW (23,000shp), 4 screws

Armament: 10 305mm (12in) guns, 27 12-pounder guns, 5 457mm (18in) torpedo tubes (submerged)

Armour: Belt 279–178mm (11–7in), Bulkhead 203mm (8in), Barbettes 279–102mm (11–4in), Turrets: 279mm (11in), Conning tower 279–203mm (11–8in)

Range: 12,260km (6620nm) at 10 knots (18.5km/h/11.5mph)

Speed: 21.6 knots (40km/h, 25mph)

Complement: 773

Cleared for action – note the collapsible railings which might otherwise be blown away by the blast. The 12-pounder guns and sighting hoods on the 305mm (12in) turret are also clearly visible.

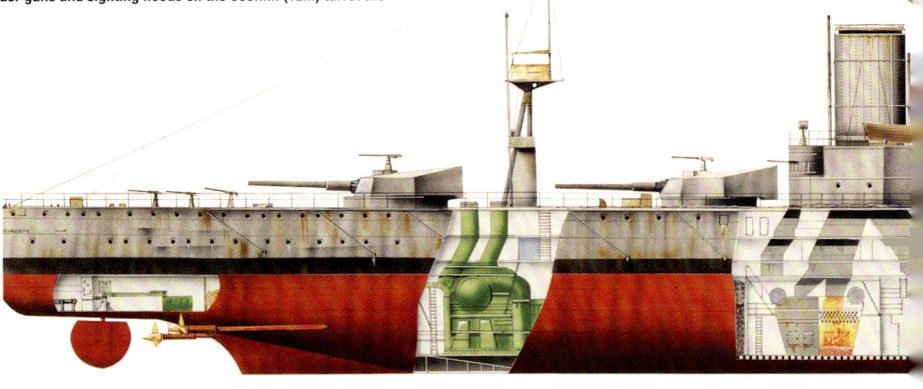

WORLD WAR I

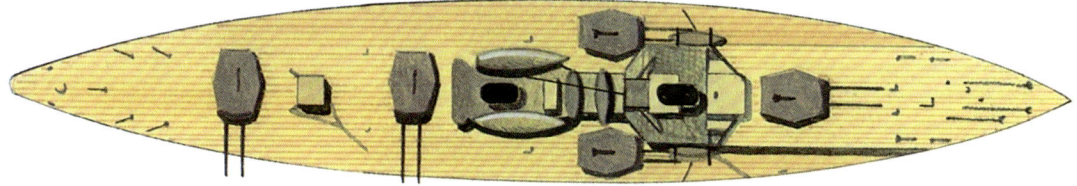

speed. The Italian naval engineer Vittorio Cuniberti had advocated the 'all-big-gun' ship in 1903 in an article titled 'An Ideal Battleship for the British Fleet', and the concept had been much discussed since. Turbine propulsion for a major warship was something new, but turbines were already in use on merchant ships, and the new fast Atlantic liners *Lusitania* and *Mauretania* were turbine-driven. The Vickers Mk X 305mm (12in) guns were a proven design. However, only *Dreadnought* had all these features. It did not entirely set the pattern for the future: USS *South Carolina*, with its superfiring turrets, was already being built when *Dreadnought* was launched, and the 356mm (14in) gun was already under development.

Prototype ship

Much of the credit for the design was claimed, and accorded to, Britain's First Sea Lord, Admiral Sir John Fisher, but the members of his Committee on Naval Design, formed in 1904, made important contributions. They

Foremast

Dreadnought was the first battleship to have a tripod foremast, giving greater stability to the traditional pole type and reducing the need for wire stays.

Main guns

The layout of the main guns was influenced by Admiral Fisher, who considered that the ability to fire forward (or rearwards) was more important than the mounting of a full broadside.

appreciated that a battleship could now be so large, so fast and capable of firing at such a long range, that the nature of naval battle would change. They also knew that current work on fire control systems, state-of-the-art on *Dreadnought*, would help bring that change about. The importance of *Dreadnought* is as a prototype of the ships that would fight the next war, by which time there would be super-dreadnoughts, with *Dreadnought* itself reduced to secondary status.

The hull design was conventional, with a bow of reduced ram form, a forecastle and foredeck on which the front turret was mounted and a main deck extending from the deckhouse to the stern. Its internal

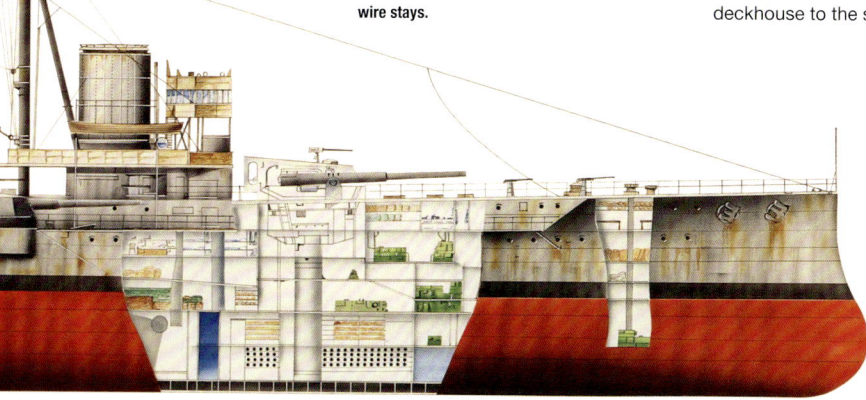

19

WORLD WAR I

Though *Dreadnought* had the latest fire control systems, the limited extent of these is clear by comparison with the superstructures of post-1915 capital ships.

arrangement included 19 watertight bulkheads and watertight fire-doors between compartments. A significant external difference was the absence of a casemate mounting a secondary armament; instead a 305mm (12in) gun turret was mounted on each side of the tripod foremast, whose support legs were set forward. Unusually, this mast was set behind the forward funnel. The mainmast was a short tripod supporting a spotting and gun control platform. Crew accommodation, uniquely in a Royal Navy vessel, was described as 'lavish and spacious': the men were housed in the after part, traditionally the officers' domain, whereas the officers, although given large cabins, were placed forward, close to their work positions.

Unmatched speed

The machinery consisted of 18 Babcock & Wilcox 3-drum mixed-fuel boilers, with a working pressure of 17.6kg/sq cm (250psi), in three boiler rooms, feeding steam to two sets of two Parsons direct-drive turbines, high- and low-pressure, ahead/astern respectively, driving four shafts. Total shaft power output was 17,000kW (23,000shp), intended to give a top

speed of 21 knots: on sea trials the ship attained 21.6 knots (40km/h, 24.9mph). Bunker capacity was 2914 tonnes (2868 tons) of coal and 1140 tonnes (1120 tons) of fuel oil. Maximum range at 10 knots (18.5km/h, 11mph) was 12,260km (7620mi, 6620nm). On trials, *Dreadnought* steamed 11,200km (7000mi) at an average 32.2km/h (20.3mph, 17.5 knots), a feat unachievable by any other warship.

Electric power was supplied from three 100kW DC Siemens generators, initially driven by two Brotherhood radial steam engines and two Mirrlees diesel engines – among the earliest to be installed on a naval vessel. Apart from the communications system, electricity powered the ammunition hoists, winches, ventilation fans and ship's lighting.

Up-to-date fire control was installed from the start and modernized several times during the ship's career. Initially, Barr & Stroud optical rangefinders, 2.7m (9ft) wide were installed in the masthead control stations. Data was passed to a fire control station and a Dumaresq mechanical computing device mounted in the conning tower, which transmitted directions electrically to the turrets. *Dreadnought* was also fitted with a Vickers range clock that could continuously calculate the distance between the ship and its moving target. A torpedo control station was mounted in the lower crosstrees of the foremast. In the course of a refit in 1912–13 the system was comprehensively updated.

Gunpower

The guns could elevate to 13.5° and fired 362kg (800lb) armour-piercing (AP) shells with a muzzle velocity of 831m (2725ft) per second, to a range of 15,040m (16,450yds). The range could be extended with heavier 390kg

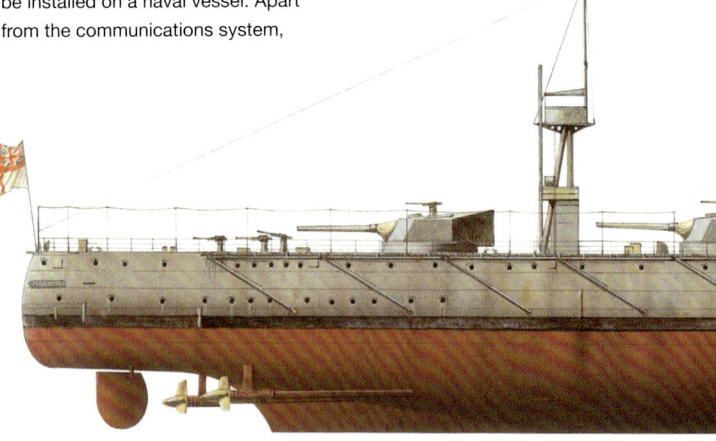

(850lb) shells with an increased crh (calibre radius head) that gave a flight of 17,240m (18,850yds). A six-gun salvo could be fired forwards and an eight-gun lateral salvo (between 60° before the beam and 50° abaft the beam). One of the advantages of a salvo fired from guns all of the same calibre was that spotters could clearly identify the splashes of shells falling in the water, without having to work out or guess which battery they were from, and immediately inform all turrets. Firing rate was two rounds per minute, and a total of 800 rounds were carried in the magazines. Around 22 torpedoes were also carried, fired through lateral tubes on each side and a stern tube.

'All big gun' did not mean no other guns, simply the omission of medium calibre and range guns. *Dreadnought* originally carried 27 76mm (3in) QF guns, primarily to fight off torpedo boats or other fast, light vessels at close range. They had a rate of fire of 20 rounds a minute.

Krupp cemented armour (KCA) was fitted externally. Internal protection included 200mm (8in) shatter-resistant mild steel for the voice pipe from the conning deck to the transmitting station on the armoured middle deck. 50mm (2in) torpedo bulkheads protected the magazines of A, X and Y turrets, with double thickness on the more exposed lateral P and Q turrets.

Dreadnought was flagship of the Home Fleet from 1907 to 1911. In 1914, it was briefly flagship of the 4th Battle Squadron at Scapa Flow. On 18 March 1915 it rammed and sank U-29, the only known occasion of a surface ship sinking a submarine in this way. Decommissioned in 1920, it was scrapped in 1921.

Anti-torpedo nets

Anti-torpedo nets were still a standard feature, and are seen here both as deployed and with the booms (first wood, later steel) shipped against the side.

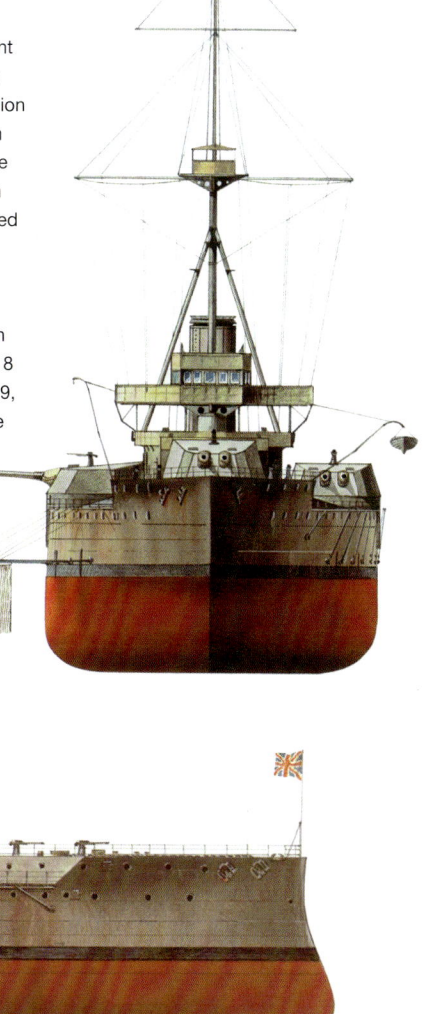

WORLD WAR I

Schleswig-Holstein (1908)

GERMANY

The only pre-dreadnoughts to see action in both world wars were *Schlesien* and *Schleswig-Holstein* of the *Deutschland* class. Commissioned at Kiel on 6 July 1908 as the last of its class, *Schleswig-Holstein* exemplified the state battleship design had reached before the 'all-big-gun' concept was established.

Design of the class was in 1902–03, although modifications were made to the later ships. The name was a provocative one: *Schleswig-Holstein* was a province gained by Germany from Denmark after two nineteenth-century wars (the crew called the ship *Sophie X*). Its dimensions were required to fit the current width of the Kiel Canal, to enable rapid transfer from the Baltic to the North Sea. In its original form, it had three funnels and two pole masts, with dismountable topmasts.

Action in World War I

Despite its relatively light armour and slow speed, *Schleswig-Holstein* was active in the North Sea during World War I, including the Battle of Jutland, where it fired around 20 rounds, and took a hit from a 381mm (15in) shell. Repaired, but decommissioned in 1917, it was used as a floating barracks, but recommissioned in 1926.

Modifications made in the inter-war period included the trunking of the fore-funnel into the second. The superstructure was extended aft to provide extra accommodation in its role as flagship (1926–35). An admiral's observation/command bridge was fitted, the navigation bridge enlarged, and searchlight platforms were mounted on the foremast. Fire control stations were positioned behind the foremast and at the base of the mainmast, with a rangefinder and spotting platform topping the foremast. The underwater torpedo tubes were removed and replaced by four 500mm (18in) torpedo tubes at casemate level

(removed in 1935). Eight boilers were converted to oil firing but four remained coal-fired, which became useful in 1944–45 when oil was in short supply.

The 280mm (11in) guns were retained throughout the ship's career, but by 1939, the secondary armament consisted of six 105mm (4.1in), four 37mm (1.45in), and four 20mm (0.8in) flak guns. By 1944 the AA armament was supplemented with 10 40mm (1.6in) Bofors guns and 22 additional

The ship in its post-1926 form, steaming off Świnoujście (Swinemünde) on the Baltic coast.

20mm (0.8in) guns, in anticipation of use as a convoy escort, which never happened.

Schleswig-Holstein is remembered as the ship that fired the opening shots of World War II, at Danzig (Gdansk) on 1 September 1939, when Germany invaded Poland. Afterwards it was mostly used as a training ship for cadets. Following RAF air attacks in December 1944, it sank at Gotenhafen (present-day Gdynia).

Refloated by the Soviet Navy, it was towed into the Gulf of Finland, and used for target practice until the 1960s.

Schleswig-Holstein

Dimensions: Length 127.6m (418ft 8in), Beam 22.2m (72ft 10in), Draught 8.2m (26ft 11in), Displacement: 13,190 tonnes (12,982 tons); 14,218 tonnes (13,993 tons) full load

Propulsion: 12 water-tube boilers, 3 vertical triple-expansion engines, 3 screws, 13,000kW (17,000shp)

Armament: 4 280mm (11in) guns, 14 170mm (6.7in) guns, 14 150mm (5.9in) guns, 4 machine guns, 6 450mm (17.7in) torpedo tubes

Armour: Belt 240–100mm (9.4–3.9in), Deck 97–40mm (3.8–1.6in), Turrets 280mm (9.4in) max, Conning tower 300mm (11.8in)

Range: 10,590km (5720nm) at 10 knots (18.5km/h/11.5mph)

Speed: 19.1 knots (35km/h, 21.8mph)

Complement: 771

Funnels

In its original form, the ship had tall upper masts and three funnels. This is a post-1926 view.

Schleswig-Holstein making smoke in the course of manoeuvres.

WORLD WAR I

Indomitable (1908)

UNITED KINGDOM

Fast armoured cruisers had long featured in fleets but they were relatively small, around 10,000 tonnes (9842 tons), and typically armed with guns of up to 228mm (9in).

Warships were becoming larger in the 1900s, and Britain's ever-inventive Admiral Fisher sought a class of ship that would be a world-beater like *Dreadnought*, combining the speed of a cruiser with the armament of a battleship. The *Invincible* class, of which *Indomitable* was the second to enter service (June 1908), was the result. It was actually 1911 before the term 'battlecruiser' was officially bestowed on them and certain ships that followed them (see HMS *Lion*). The major navies watched one another very closely, both officially through diplomatic naval attachés and unofficially through open observation and covert spying. By 1911, battlecruiser squadrons were a significant addition to other fleets.

The long hull (to improve speed) and low superstructure of the class were to typify the look of battlecruisers. Their laden displacement of over 21,055 tonnes (20,722 tons) was primarily due to the heavy machinery required for speed, and to the weight of their four 305mm (12in) twin gun turrets.

Need for speed

Speed was one of the desired qualities. *Indomitable* got steam from 31 Babcock & Wilcox water-tube boilers, fitted with oil sprayers to increase burning heat. It was driven by two paired sets of Parsons direct-drive turbines, driving four propellers and designed to provide 31,000kW (41,000shp) but substantially exceeding this in trials, achieving a top speed of 48.3km/h (30mph/28.1knots) with a power output of almost 36,000kW (48,000shp). Even its designated speed of 47.2km/h (29.3mph, 25.5nm) was faster than the preceding armoured cruisers of the *Minotaur* class (1906).

Large coal consumption was to be expected, and bunker capacity was 3132 tonnes (3093 tons) of coal. Cruisers were expected to have a long operating range, and in this respect

Indomitable was not exceptional, with a maximum of 5720km (3090nm) at 10 knots.

Hefty armament

The other desired quality was a substantial armament, enough to destroy any other warship short of a battleship. Eight 305mm (12in) guns in four twin turrets were fitted (Minotaurs had four 234mm (9.2in) and ten 191mm (7.5in) guns). Guns of this calibre had hitherto been restricted to battleships, and information about the new class's armament had been carefully guarded. Five underwater torpedo tubes were fitted, two each side and one firing through the stern.

Armour, as can be seen from the specifications, was on the meagre side for a ship that might be expected to face enemies of similar firepower. It did, however, include internal anti-torpedo bulkheads to shield the magazines and shell rooms.

In the Battle of Dogger Bank on 23 January 1915, *Indomitable* was one of the ships that sank the German heavy cruiser *Blücher*. It also towed the heavily damaged HMS *Lion* back to port. It was engaged in the Battle of Jutland, without damage, whereas three other British battlecruisers were destroyed.

Armour

The length of hull required for higher speeds resulted in a thinner spread of armour in order to keep down overall weight, sacrificing protection for speed.

HMS Indomitable

Dimensions: Length 173m (567ft), Beam 23.8m (78ft), Draught 7.3m (27ft)
Displacement: 17,250 tonnes (17,525 tons); 21,055 tonnes (20,722 tons) full load
Propulsion: 31 Babcock & Wilcox water-tube boilers, two paired sets of Parsons direct-drive steam turbines, 4 screws, 31,000kW (41,000shp)
Armament: 8 305mm (12in) guns in twin turrets, 16 102mm (4in) guns in single mounts, 5 450mm (18in) torpedo tubes
Armour: Belt 152–102mm (6–4in), Deck 64–38mm (2.5–1.5in), Barbettes and Turrets 178mm (7in), Conning tower 254–152mm (10–6in)
Range: 5720km (3090nm)
Speed: 25.5 knots (47.2km/h, 29.3mph)
Complement: 784

WORLD WAR I
Vittorio Emanuele (1908)

ITALY

Naval architects and naval administrators mixed like oil and water, especially if the designer was a man of unusual talent and foresight.

So it was that Vittorio Cuniberti (1854–1913), chief constructor of the Regia Marina (Royal Navy) and the first person to propose the 'all-big-gun' fast battleship in print (1903), failed to get clearance to build the 12-gun 'colossus' he had proposed. Instead, Cuniberti had to settle for a design that had only two 305mm (12in) guns as its main armament. The slow pace of Italian construction and completion meant that the Regina Elena class, of which this was the second member, took eight years from design in 1900 to completion in August 1908 (see HMS Dreadnought).

Superior speed
Built by the Castellammare di Stabia naval shipyard, it had two 305mm (12in) guns in single turrets as its main armament. Compared to France's similarly dimensioned République class of 1906, with four 305mm (12in) guns, Vittorio Emanuele was under-gunned and under-armoured. In speed, however, it was superior even to Dreadnought, driven by two vertical four-cylinder triple-expansion engines, with steam from 28 Belleville water-tube boilers. It was acknowledged by the Italian Ministry of Marine, after the Battle of Tsushima, that the Regina Elena class would not be able to maintain battle against modern capital ships.

Three tall funnels
Its appearance was dominated by three tall funnels set between the masts, on a long, low superstructure flanked by casemates mounting the 76mm (3in) guns, with the 200mm (8in) guns set above on the main deck which extended aft to the mainmast. The armour was of KC (Krupp cemented) steel fabricated under licence in Italy.

Fire control was provided by Barr & Stroud rangefinders, of the type that

Vittorio Emanuele
Dimensions: Length 144.6m (474ft), Beam 22.4m (73ft), Draught 8.58m (28ft 1in), Displacement 14,137 tonnes (13,914 tons)
Propulsion: 28 Belleville boilers, 2 vertical triple-expansion engines, 2 screws, 14,484kW (19,424shp)
Armament: 2 305mm (12in) guns in single turrets, 12 200mm (8in) guns, 16 76mm (3in) guns, 2 450mm (18in) torpedo tubes
Armour: Belt 250mm (9.8in), Turrets 200mm (8in), Deck 38mm (1.5in), Conning tower 250mm (10in)
Range: 19,000km (10,000nm) at 10 knots (18.5km/h/11.5mph)
Speed: 21.36 knots (39.56km/h, 24.58mph)
Complement: 764

WORLD WAR I

imposed two images over each other, mounted on the conning tower. Little in the way of modernization happened in the course of its career. Sister ships *Roma* and *Napoli* had their foremasts removed and funnels modified, but *Vittorio Emanuele* retained its original appearance. *Vittorio Emanuele* was involved in the Italian–Turkish War of 1911, engaging mostly in shore bombardment. When Italy entered World War I against the Central Powers in 1915, its role was confined to patrolling the southern end of the Adriatic Sea, to keep the ships of the Austro-Hungarian empire bottled up. It was stricken on 1 April 1923 and broken up.

Coal furnaces produced dense smoke. The tall funnels lifted the smoke high above the decks and also helped to create an up-draught to maintain the fires.

Main guns
The two main guns were of British-designed Elswick 'I' pattern, with electric mountings, 200 elevation and a range of 20,000m (21,870yd).

WORLD WAR I

South Carolina (1910)

UNITED STATES

This was the first warship to have superfiring turrets. Numbered BB-26, this was a first class dreadnought battleship and class leader, although USS *Michigan* (BB-27) was actually commissioned before it.

Laid down at Cramp's Yard, Philadelphia on 18 December 1906, it was launched on 11 July 1908 and commissioned on 1 March 1910. The total cost was $7,000,000. Despite the 'dreadnought' appellation, it was designed before HMS *Dreadnought*, although not completed until four years after the British ship, and it did not have turbine drive. It was fitted with woven wire masts, holding stations for the fire control officers who had telephone contact with the gun turrets.

First lattice masts

BB-26 and 27 were the first US ships to carry the so-called lattice masts, often referred to as basket masts and

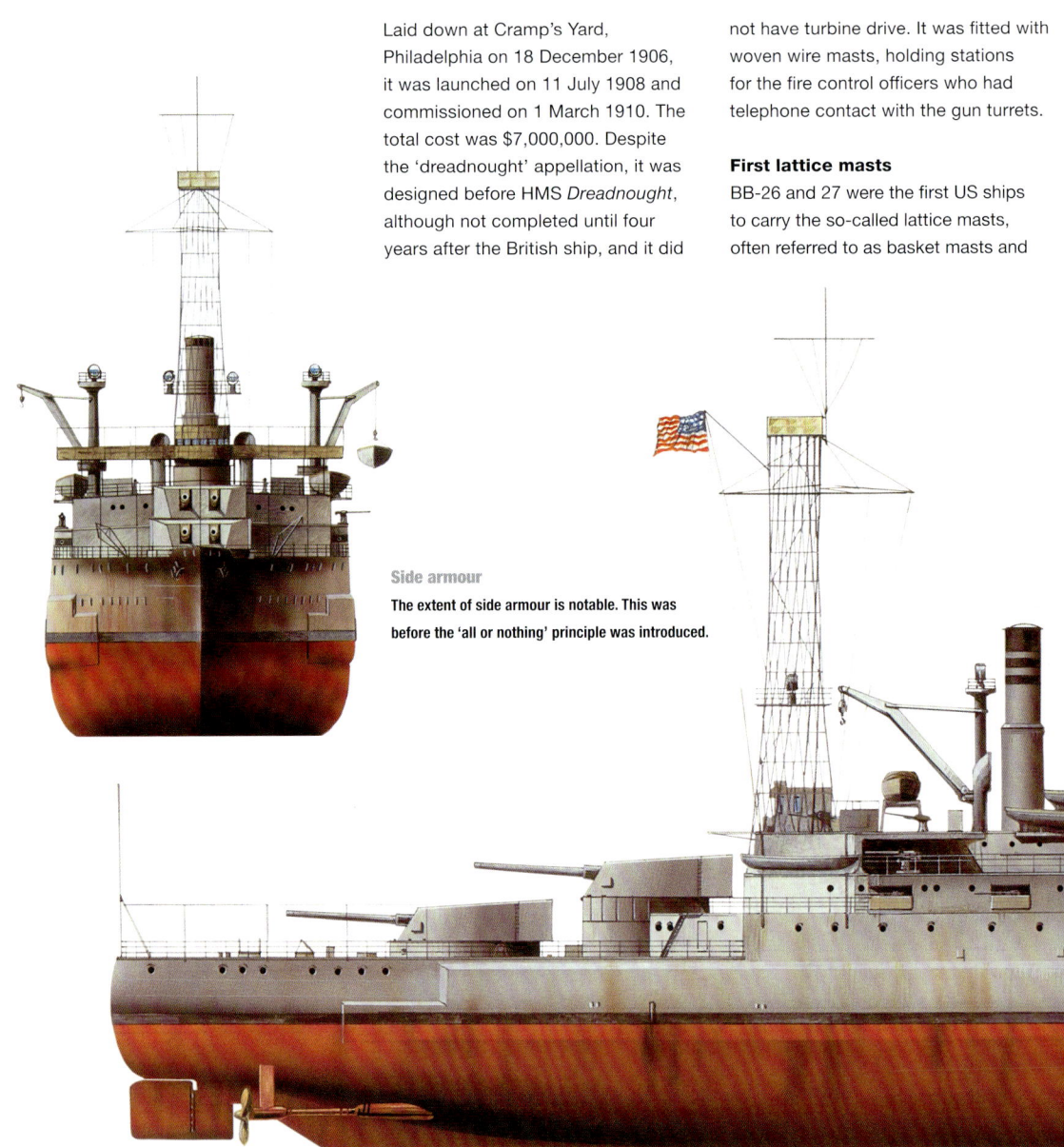

Side armour
The extent of side armour is notable. This was before the 'all or nothing' principle was introduced.

unfavourably compared to electricity pylons, which were to typify US battleships until the 1930s. In fact, they were multitubular steel towers in which the topmost and bottom retaining rings were turned in opposite directions, resulting in a rotary hyperboloid form. It was felt that the nature of the structure would protect fire control positions from the shocks and vibrations that were transmitted through the traditional form of mast. Tested in gunnery practice, the structure proved stable against shell hits, although *Michigan's* forward mast was brought down in a gale in 1918. The top platform was exposed and unsuitable as a control post or gun platform, and the design could not support an armoured fire-control position, so the basket masts were eventually replaced as ships of the period were modernized. Decorative heraldic-type swags and wing-forms were fitted to the bow but the 'fanciwork' was discarded by March 1910.

Superfiring system

The most distinctive and significant feature, however, was the superfiring turrets, which enabled independent fire from four big guns. It also meant that all the main armament could be placed on the centre-line. As far back as 1903, US naval designers had considered the 'all-big-gun' battleship, with 305mm (12in) guns and 76mm (3in) guns to deter torpedo boats. Such a ship would have the ability to fire heavy shells well beyond the range of contemporary torpedoes. By 1905, the issue moved from theory to demand. The Navy's Board on Construction wanted big-gun ships. US Congress set the size limit at 16,256 tonnes

---

USS South Carolina

Dimensions: Length 137.9m (452ft 8in), Beam 24.5m (80ft 5in), Draught 7.49m (24ft 7in), Displacement 16,256 tonnes (16,000 tons); 18,186 tonnes (17,900 tons) full load

Propulsion: 12 Babcock & Wilcox boilers, 2 vertical 4-cylinder triple-acting engines developing 12,304kW (16,500shp), 2 screws

Armament: 8 305mm (12in) guns, 22 88mm (3in) QF guns, 2 underwater 533mm (21in) torpedo tubes

Armour: Belt 305–228mm (12–9in), Casemates and barbettes 254–203mm (10–8in), Turrets 304mm (12in), Deck 38mm (1.5in), Conning tower 305mm (12in)

Range: 9260km (5000nm) at 10 knots (18.5km/h, 11.5mph)

Speed: 18.5 knots (34.3km/h)

Complement: 806

WORLD WAR I

(16,000 tons). It was at this time that the superfiring turret came under active consideration. With the bodies responsible for gunnery, machinery and construction acting in concert (which had not always been the case), a design emerged in which eight big guns, along with powerful machinery and adequate armour, could fit within a 16,000 ton limit and fire the same weight of broadside as the 10-gun *Dreadnought*.

Criticism of turrets

Other navies were quick to emulate the superfiring system and the pattern for most subsequent capital ships was set, with two turrets forward and aft of the main superstructure. It was not achieved without controversy: in some quarters, the superfiring turrets were criticized for providing a 'single target' for enemy guns. The man behind the project was Rear Admiral Washington L. Capps, head of the Bureau of Construction and Repair. Capps was also responsible for novel internal arrangements, brought about by the need to create central space and keep down weight. In previous ships, the armoured deck sloped down towards the sides, meeting the top of the external armour plating. This was replaced by a flat deck at the level of the armoured belt's upper edge and the belt itself was thickened at the vital points to protect machinery and magazines, between the foremost and aftmost barbettes.

Crew space was largely at the stern for officers and warrant officers; the seamen slung their hammocks in crew spaces around the forward barbettes, above the magazines. US ships normally took more account of their crew's comfort than those of the European and Japanese navies: *South Carolina* had a firemen's washroom. In 1918 the large charthouse at the base of the mainmast was removed and the conning tower enlarged, with a small fire control tower at the aft end.

Twelve Babcock & Wilcox boilers were fitted, working at 20.65kg/sq cm (295psi) and feeding steam to two vertical inverted, direct acting 4-cylinder triple-expansion engines generating 12,304kW (16,500shp) at 125 revolutions per minute. Lateral coal bunkers ran along each side of the boiler rooms, with a total capacity of 2133 tonnes (2100 tons). Coaling at sea tests were run in 1913 and 1914, with 363kg (800lb) bags slung across from the collier *Jupiter*. The process was considered too slow to be of practical

South Carolina around 1918, with range-finders mounted on the turrets and topmasts heightened for long-range radio communication.

use, and dropped. The torpedo rooms were between the forward barbette and the ship's bows. *South Carolina* also carried 18 contact mines, designed to hang 1.4m (5ft) below the surface. The gun turrets were provided with sighting hoods on the sides, to avoid blast damage from superfire to roof-mounted hoods.

Boats carried were three 12.2m (40ft) steamers, two 11m (36ft) motor sailing launches, one 9.4m (31ft) racing cutter, two cutters, two 9.1m (30ft) whaleboats, two 6m (20ft) dinghies, four 6m (20ft) punts. The total capacity of these boats was 433 men, just over half the total crew number of 806. Hospital accommodation was forward of the main battery. *South Carolina*'s last task was to be used for testing torpedo blisters at Philadelphia in May 1924. After that the ship was scrapped.

WORLD WAR I

Hercules (1911)

UNITED KINGDOM

When *Hercules* was laid down on 30 July 1909, British designers were being required to maximize the broadside capacity of battleships.

Main guns
The 305mm (12in) guns were Vickers Mk XI 50-calibre, the long barrels intended to increase muzzle velocity and range. *Hercules* fired 98 rounds at Jutland.

The three ships of the *Colossus* class had two 305mm (12in) turrets centre-mounted *en echelon*, enabling all 10 guns to be fired to either port or starboard. This required the creation of space amidships, with the P and Q turrets separated by the after funnel. Flying decks were constructed both to hold the ship's boats and to provide a passageway.

Several alterations were made before the ship saw action: a searchlight platform was installed, the fore funnel was heightened and a 'clinker cowl' fitted, to stop or reduce the amount of ash that fell on the navigating bridge. The persistence of placing this funnel so near the bridge was inconvenient, but forced by the position of the boiler rooms below. Wartime alterations included the removal of the flying bridges, as they were considered too exposed. Unusually for a Royal Navy battleship, a single mast was fitted.

Coal-fired by Yarrow
Hercules was coal-fired, with 18 Yarrow boilers working at 16.1kg/sq cm (230psi), serving 4 Parsons direct-drive steam turbines. Oil sprayers were fitted to speed up the process of raising steam. The inner high-pressure turbines drove the inner propellers and low-pressure turbines drove the outer propellers. On trial the engines developed 19,835kW (26,599shp) at 345rpm, with a speed of 38.5km/h (24mph, 21 knots). There were three engine rooms, with the two inner shafts operated from a central room and each outer shaft from separate rooms to port and starboard. Bunker capacity was 2946 tonnes (2900 tons) of coal and 813 tonnes (800 tons) of oil.

Testing new devices
In late 1912, this was one of 12 ships fitted with short distance 8-km (5-mile) radio. Gun direction equipment was installed in 1915. *Hercules* was also among the ships trying out new devices such as the Waymouth-Cooke rangefinder, a handheld instrument combining a sextant with a telescope, which was used for keeping a ship in station with the rest of a squadron as well as for fire control.

A variety of range-finding and fire-control instruments were tested on ships throughout the war as makers tried to make effective devices. Another set tested on *Hercules* was 'Walker's

WORLD WAR I

Hercules

Dimensions: Length 166.3m (545ft 9in), Beam 26m (85ft 2in), Draught 8.2m (27ft), Displacement: 20,350 tonnes (20,030 tons)
Propulsion: 18 Yarrow boilers, 4 sets Parsons direct-drive steam turbines, 2 screws, 19,000kW (26,000shp)
Armament: 10 305mm (12in) guns in twin turrets, 16 102mm (4in) guns in single mounts, 3 533mm (21in) torpedo tubes
Armour: Belt 279–203mm (11–8in), Turrets 280mm (11in), Barbettes 279–102mm (11–4in), Conning tower 280mm (11in)
Range: 12,370km (6680nm) at 10 knots (18.5km/h, 11.5mph)
Speed: 21 knots (38.5km/h, 24mph)
Complement: 791

Vice-Admiral Sir Doveton Sturdee, Flag Officer of the 4th Battle Squadron, on the quarterdeck of HMS *Hercules*, c. 1916–17.

instruments' – a triple set intended to establish the deflection rate for a torpedo director with the ship in motion: a complex device that did not yield quick results in a changing situation.

In the Battle of Jutland, *Hercules* fired 98 305mm (12in) shells at enemy ships and claimed hits on the *Seydlitz*. It was sold for scrapping in 1921.

WORLD WAR I

Lion (1912)

UNITED KINGDOM

First of a new class of battlecruisers, and the first RN warship to cost over £2 million, *Lion* was a developed version of the *Indefatigable* class, faster by 3.7km/h (2 knots) and carrying eight 343mm (13.5in) guns.

On completion, it was the largest warship in the world. Initially the foremost funnel was placed ahead of the foremast, but this made the control top uninhabitable when steaming hard, and the positions were switched.

Speed versus armour

The first RN battlecruiser to have superfiring turrets, it was designed to outrun and outshoot its German equivalents *Moltke* and *Goeben* (1909–11), whose main armament was 10 280mm (11in), but its armour was somewhat lighter than theirs. This was a typical difference between British and German battlecruisers, and disadvantageous to the British. The German designers chose lighter guns in order to fit heavier armour, whereas the Royal Navy followed Admiral Fisher's dictum that speed was of greater value than armour. In a fleet action, such as Jutland, the German ships had better protection, including more internal compartments separated by watertight bulkheads, enabling them to withstand a heavy battering.

Yet the worst of the British ships' vulnerability was of their own making. Intensive competition in gunnery practice to achieve the most rapid fire led to the habit of storing shells and cordite in unprotected spaces, and of stripping out anti-flash protection to get the shells hoisted faster. If a shell should penetrate the 229mm (9in) turret armour, the resultant fire could spread downwards unimpeded, with catastrophic consequences.

Seven boiler rooms each held six Yarrow water-tube boilers: one reason for the ship's exceptional length. These were coal-fired with oil-sprayers. Total bunker capacity was 3556 tonnes (3500 tons) of coal and 1153 tonnes (1135 tons) of oil.

Fully loaded, *Lion*'s range was 10,390km (5610nm), at 18.5km/h (11.5mph, 10 knots), more than enough for a ship essentially intended to fight ships of the German fleet in the North

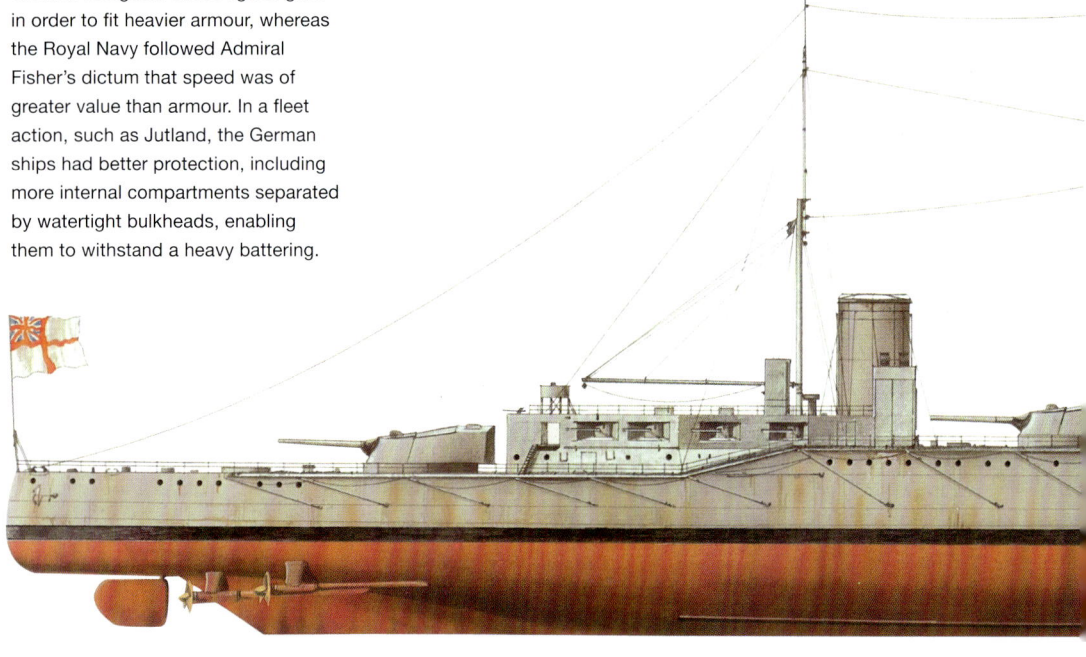

Sea, but not generous for long-range cruising. The 343mm (13.5in) guns were made by Vickers. Their weight and size made it essential for them to be on the ship's centre-line, rather than en echelon as with Indomitable.

The turrets, A,B,Q and Y were operated hydraulically and the guns could be elevated to 20°, although this exceeded the capacity of the original rangefinders, limited to 15° 21' until just before Jutland, which then enabled a range of 21,780m (23,8320yds). The rate of fire was slightly under two rounds per minute. Lion's magazines were loaded with 880 567kg (1250lb) armour-piercing shells.

In the course of the war AA guns were fitted. In early 1918, flying-off platforms were installed on Q and X platforms, with Sopwith 2F Camel, Sopwith Pup and Sopwith Strutter aircraft at different times.

Flagship at Dogger Bank

Lion was Admiral Beatty's flagship in the Dogger Bank engagement on 24 January 1915. While the British tactic was to fire at all enemy vessels as soon as possible, the Germans concentrated their initial fire on the leading enemy ship, with the intention of breaking its ability to communicate. A 280mm (11in) shell from Moltke pierced the armour belt, causing serious damage.

HMS Lion
Dimensions: Length 213.4m (700ft), Beam 27m (88ft 6in), Draught 9.9m (32ft 5in, Displacement 26,925 tonnes (29,680 tons) full load
Propulsion: 42 Yarrow water-tube boilers, 2 sets of Parsons direct-drive steam turbines, 4 screws, 52,199kW (70,000shp)
Armament: 8 343mm (13.5in) guns, 16 102mm (4in) guns, 2 533mm (21in) torpedo tubes
Armour: Belt 229–102mm (9–4in), Bulkheads 102mm (4in), Barbettes 229–203mm (9–8in), Turrets 229mm (9in), Deck 64–25mm (2.5–1in), Conning tower 254mm (10in)
Range: 10,390km (5610nm) at 10 knots (18.5km/h, 11.5mph)
Speed: 28 knots (51.9km/h, 33.4mph)
Complement: 997

Early profile

The image shows the original appearance of the ship. The main guns are Vickers BL Mark V 343mm (13.5in) 45 calibre: much more effective than the 305mm (12in) guns of the Colossus class.

WORLD WAR I

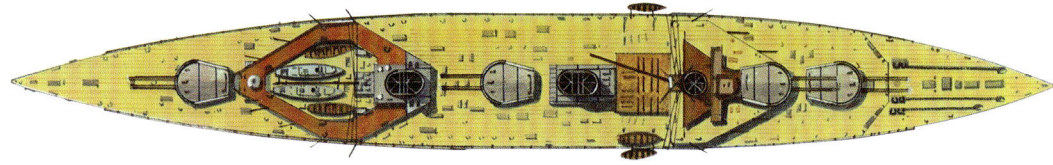

Deck plan
The original deck plan and profile display the elegant form of the hull.

Lion took heavy fire also from *Seydlitz* and *Derfflinger*, suffering 16 shell strikes. It lost all power and had to be towed back to Rosyth by HMS *Indomitable*.

With the repairs after the Battle of Dogger Bank, *Lion* was fitted in early 1915 with a director firing system. The fire control director was mounted high in the superstructure and fed data directly to the turrets via electric wires. From the control position all main guns could be fired simultaneously. A second director was added in 1918.

What the Royal Navy did not do in 1915 (unlike the Imperial Navy) was revise its procedures of ammunition and cordite storage below the gun turrets and install heavy doors and hatches to prevent 'flash fire' reaching downwards and into the magazines. This failure would lead to the loss through flash fire of three battlecruisers at Jutland: *Invincible*, *Indefatigable* and *Queen Mary*, and almost *Lion*.

Performance in battle

In battle, spotters in the control top, with powerful binoculars, noted the shell splashes and hits, and the range takers observed the convergence or divergence of the enemy line. The range-takers in the range-finding tower had high-powered optics with a narrow field of view. All passed information to the windowless transmitting station where computation of bearing and elevation of the guns was worked out and sent to the turrets and director. From here, the guns were fired at a point when the ship, if in a rolling sea, was stable.

In the Battle of Jutland, with *Lion* again flagship of the battlecruiser squadron, a direct hit on Q turret by a 305mm (12in) shell from *Lützow* blew its 229mm (9in) armoured roof off and killed everyone inside. Despite having his legs blown off, Major Harvey of the Royal Marines ordered the magazine flood valve to be opened before the resultant fire spread downwards, saving the ship and winning a posthumous Victoria Cross.

Lion was placed on reserve in 1920 and scrapped in 1924.

These images of *Lion* show the post-1912 appearance, with heightened funnels and repositioned foremast.

WORLD WAR I

Wyoming (1912)

UNITED STATES

Wyoming (BB-32) was ordered in 1909, laid down at Cramp's Yard in Philadelphia in 1910 and completed in 1912.

In response to changing requirements in technology and communications, a battleship's profiles might change considerably during the ship's life. This was certainly true of *Wyoming*. In its original form it mounted no less than six twin 305mm (12in) turrets, numbered I to VI, four of them pointing aft, on a long afterdeck. The arrangement was used only on *Wyoming* and sister ship *Arkansas*. Although the ships were considered *dreadnoughts*, they were not 'all-big-gun' like the European kind. No less than 21 127mm (5in) guns were carried, mostly in lateral casemates on an extended forecastle.

In World War I *Wyoming* was one of the four ships in BatDiv9, sent across the Atlantic to join the British Grand Fleet during 1917–18. It was also among the ships escorting the surrendered German High Seas Fleet to Scapa Flow in November 1918.

Post World War I refits

Refits began in 1919 when some 130mm (5in) guns were removed from the lower casemate, because they were too near the waterline (not for the first time on a US battleship). In 1925–27 it was converted to oil burning with four White-Forster boilers and a single funnel, the torpedo tubes were removed and anti-torpedo bulges fitted, the armoured deck was thickened, the basket-type mainmast was replaced by a heavy three-platform tripod set further aft, and a floatplane catapult was installed on 'III' turret.

In 1933, to conform with the Washington naval treaty, the ship was partially disarmed, losing the guns of III, IV and V turrets and the fantail 152mm (6in) gun. It was reclassified as a training ship, AG-17 (AG denoted 'miscellaneous auxiliaries'), and used for this purpose until November 1941 when it became a gunnery training ship, operating mainly in the Chesapeake Bay area, primarily with AA guns,

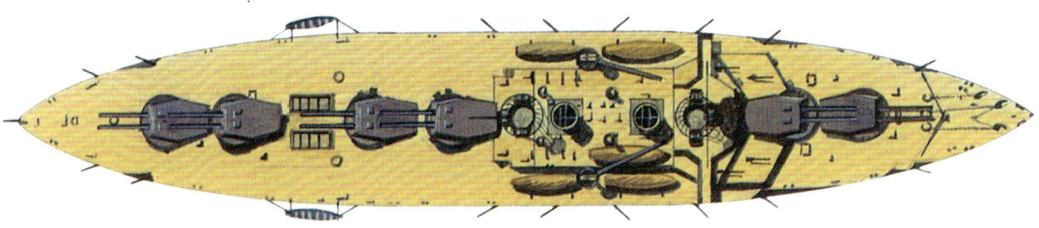

Deck plan

As the deck plan shows, *Wyoming* could mount a formidable 12-gun broadside to either port or starboard.

of which a variety of types were mounted, mostly 40mm and 20mm, in open mounts. In 1944 the side armour was removed, the basket foremast was replaced by a single pole and the remaining big guns were taken out.

It was decommissioned on 1 August 1947 and sold for scrapping in October of that year.

USS Wyoming

Dimensions: Length 171m (562ft), Beam 28.4m (93ft 2in), Draught 8.7m (28ft 6in), Displacement 26,416 tonnes (26,000 tons)
Propulsion: 12 Babcock coal/oil boilers, 4 Parsons direct-drive turbines, 4 screws, 20,880kW (28,000shp)
Armament: 12 305mm (12in) guns, 21 130mm (5in) guns, 2 533mm (21in) torpedo tubes
Armour: Bulkhead 279–229mm (11–9in), Belt 279–127mm (11–5in), Decks 76–25mm (3–2in), Turrets 305mm (12in), Barbettes 279mm (11in), Funnels 165mm (6.5in), Conning tower 305mm (12in)
Range: 14,800km (8000nm) at 10 knots (18.5km/h, 11.5mph)
Speed: 20.5 knots (37.9km/h, 23.7mph)
Complement: 1063

Main guns
The main guns were 50 calibre Mk 7, introduced for the *Wyoming* class and made at the US Naval Gun Factory. Maximum elevation was 150°; they fired 390kg (870lb) AP shells to a range of 18,000m (19,680yds).

WORLD WAR I

Viribus Unitis (1912)

AUSTRIA-HUNGARY

The Imperial Austrian Navy's base at Pola was in an unfortunate strategic position at the head of the long, narrow Adriatic Sea. Its only likely rival was Italy, but as Italy had battleships, so must Austria – hence *Viribus Unitis*.

Viribus Unitis was laid down in 1908 on the 'all-big-gun' principle, and built quite quickly at Stabilimento Tecnico Triestino: laid down on 24 July 1910, launched on 24 June 1911 and commissioned on 5 December 1912. There was a sister ship, *Tegetthoff*. The chief designer was STT's Siegfried Popper although there was much wrangling with the Naval Technical Committee. Popper was given access to plans of Germay's *Kaiser* class, but also had Italy's new *Dante Alighieri* battleships in his sights. They had 12 305mm (12in) guns in triple turrets. The *Tegetthoff* class was given the same main armament, all mounted on the centre-line and using superfiring turrets.

First triple-turret

As *Viribus* was commissioned a month before *Dante Alighieri*, the Austro-Hungarian Navy could claim the first triple-turreted *dreadnought*. The secondary batteries, 12 150mm (5.9in) guns mounted in casemates on each side, were of slightly greater calibre, but the Italian ship, of almost exactly the same displacement, had more smaller guns. A few modifications were made in 1914, including the placing of 66mm (2.6in) AA guns in open mountings on the superfiring turrets.

Vulnerable underwater

The superstucture was a simple round-fronted tower with a closed navigating deckhouse above, on a platform built around the pole foremast and partially supported by lateral lattice frames. The two funnels were set close behind and between the two very tall pole masts. Armour protection was substantial around the central parts but dwindled towards the ship's extremities. A double bottom was fitted, coming up to meet the base of the armour plate, but of relatively thin plates, the inner torpedo bulkhead being only 50mm (2in). Both this ship and *Tegetthoff* were to succumb to underwater explosives.

Main guns
The main guns were Škoda 305mm (12in) 45cal, Model K10, with a maximum elevation of 200°, firing shells of 450kg (990lb) to a maximum range of 22,000km (13,750mi) and with a claimed rate of fire of three rounds per minute.

WORLD WAR I

Tegetthoff and *Viribus* were fitted with 12 Yarrow water-tube boilers, coal-fired with oil sprayers, powering four Parsons direct steam turbines in separate engine rooms, which drove four shafts. Bunker capacity was 1844 tonnes (1815 tons) and 267.2 tonnes (263 tons) of fuel oil, and operating range was 7780km (4200nm) at 19km/h (10 knots). Maximum speed was 37.6km/h (23mph, 20.3 knots), which was not quite in the *Dreadnought* class, and less than *Dante Alighieri*'s planned 42.2km/h (26.2mph, 22.8 knots).

The ship was never to be pitted against an Italian *dreadnought*. Apart from shore bombardment against the Italian city of Ancona, and the Montenegrin coast, it saw little action in 1914–18. Following the Austrian surrender, it was briefly handed over to the new state of Yugoslavia, but was sunk by limpet mines set by Italian frogmen on 1 November 1918, inside the Pola base.

The main turrets on *Viribus Unitis* weighed 561 tonnes (552t).

Viribus Unitis
Dimensions: Length 152.2m (499.25ft), Beam 27.3m (89.7in), Draught 8.9m (29ft), Displacement 22,032 tonnes (20,000 tons); 21,89 tonnes (21,346 tons) full load
Propulsion: 12 Yarrow coal/oil boilers, 4 Parsons turbines, 4 screws, 19,700kW (26,400shp)
Armament: 12 305mm (12in) guns, 12 150mm (5.9in) guns, 18 11-pounder single guns, 4 533mm (21in) torpedo tubes
Armour: Belt 280–152mm (11–5in), Decks 25–19mm (2–1.5in), Turrets 280mm (11in), Casemates 180mm (7.1in)
Range: 7780km (4200nm) at 10 knots (18.5km/h/11.5mph)
Speed: 20.3 knots (37.6km/h, 23mph)
Complement: 1087

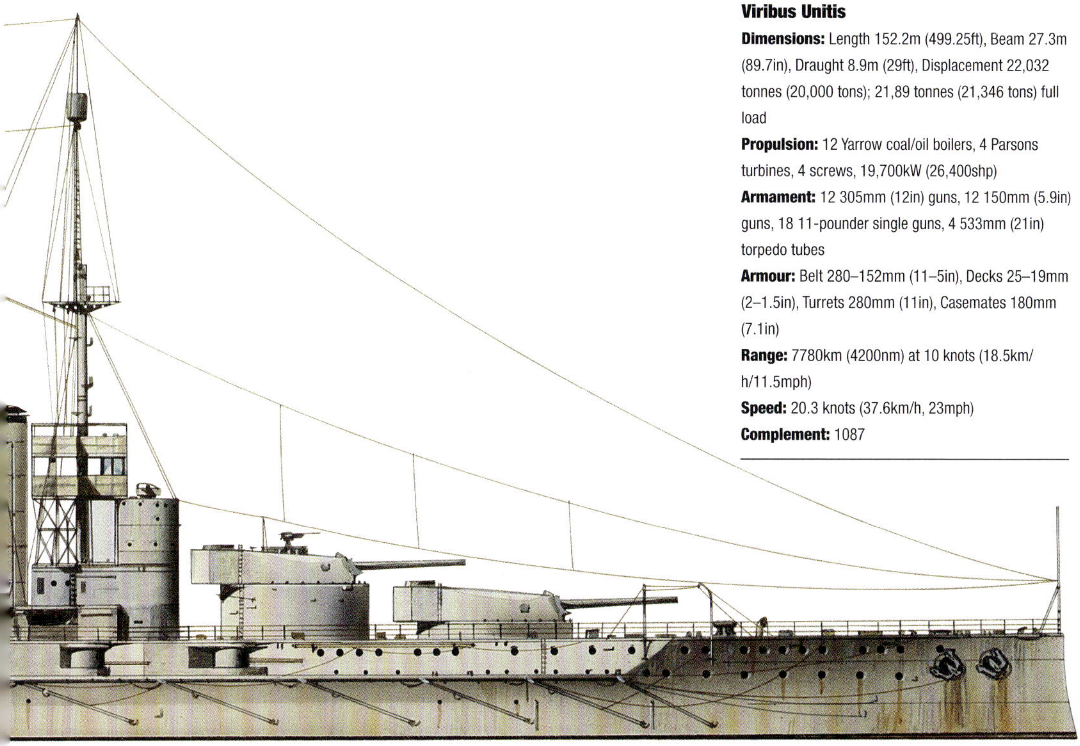

WORLD WAR I

Courbet (1913)

FRANCE

The French Navy acquired six battleships between 1909 and 1913, all of the *Danton* class and all of pre-dreadnought type.

Courbet, France's first 'dreadnought', built at the Lorient navy yard and commissioned on 19 November 1913, represented a catching-up by the fleet that a decade earlier had considered itself pre-eminent in battleship design. The designer was Léon Lyasse, Director of Naval Engineering.

Original form
In its original form it had three funnels, two of them set ahead of the foremast, which was almost double the height of the mainmast. The superstructure was low, with two bridges around a central conning tower and extending back on each side of the fore-funnel.

Twelve 305mm (12in) guns were mounted in double turrets, eight in superfiring turrets on the centre-line and four in parallel turrets to port and starboard between the second and third funnels. It did not dispense with a medium armament, having 22 138.6mm (5.4in) casemate-mounted guns.

Four beam-mounted torpedo tubes were fitted. In its original form the ship was also designed to lay 30 mines but this capacity was never put to use.

The 305mm (12in) 45 cal guns were an improved version of a 1906 design. At an elevation of 120° they fired a 440kg (970lb) shell 14,500m (15,900yds).

From 1929 the profile was changed. A 30m (98.4ft) high tripod foremast was fitted, with fire control and spotting stations high above the navigating bridge. The middle funnel was removed and the forward funnel placed on a higher mounting.

Troubled history

Its mechanical history was troubled. A serious fire in in the boiler room on 6 June 1923 necessitated substantial repairs at Toulon, and some oil-fired Guyot du Temple boilers were installed. A second fire in 1924 renewed damage in the same area. Between 1927 and 1929 a full modernization took place. Nine Guyot du Temple boilers and four Rateau-*Bretagne* turbines replaced the previous machinery, generating 32,065kW (43,000shp) but failing to improve the ship's speed. Bunker capacity was 2706 tonnes (2700 tons), oil 310 tonnes (305 tons).

Almost half the length was protected by side armour, to 2.35m (7ft 7in) above and 2.4m (7ft 9in) below the waterline, and with a maximum thickness of 270mm (10.6in), comparable with *Dreadnought*. The 12in turrets had 320mm (12.6in) armour, and the conning tower 300mm (11.8in).

In the 1924 repair, the opportunity was taken to improve the elevation of 305mm (12in) guns and so to extend the firing range. Model AA 75mm (2.9in) anti-aircraft guns were also installed.

Courbet

Dimensions: Length 168m (551ft 2in), Beam 27.9m (91ft 6in), Draught 9m (29ft 6in), Displacement 22,545 tonnes (25,850 tons); 26,4217 tonnes (26,000 tons) full load
Propulsion: 24 Belleville boilers, 4 sets of Parsons direct-drive turbines, 4 screws, 32,065kW (43,000shp)
Armament: 12 305mm (12in) guns in twin turrets, 22 138.6mm (5.4in) guns, 4 47mm (1.8in) guns, 4 450mm (18in) torpedo tubes
Armour: Belt 270–140mm (10.6–5.5in), Deck 70–40mm (2.8–1.6in), Turrets 320mm (12.6in), Conning tower 266mm (10.5in)
Range: 7800km (4200nm)
Speed: 20 knots (39km/h, 24mph)
Complement: 1085

Waterline

The waterline belt was unusually deep, at 4.75m (15ft 6in).

WORLD WAR I

Derfflinger (1913)

GERMANY

Lead ship of a class of three, *Derfflinger* was widely admired for its symmetrical proportions and handsome lines, but it was also perhaps the most effective battlecruiser design.

Battle damage

Derfflinger was hit 14 times in the Jutland battle, disabling turret A and both rear turrets, but leaving the hull intact. The British dubbed the ship 'Iron Dog' for its resistance.

The navigating bridge was set forward on the armoured deckhouse, with an admiral's bridge set back; unusually, it had a pole foremast set forward of it. Above it a pair of searchlights were mounted on the forward funnel and a rangefinder for the forward guns was also set ahead of it (a short tower aft of the mainmast held the rear guns' rangefinder).

Different hull design

The hull design showed a marked change from older practice, with a full-length flush deck. Internally it did away with transverse frames and relied on longitudinal supports, although still compartmentalized against shell or torpedo strikes: 16 watertight compartments below the waterline. This saved weight and enabled the main armour belt to be 305mm (12in) between the outer barbettes, with the same thickness in the conning tower walls (130mm, 5.1in) on the roof. In fact the degree of armour protection made the *Derfflinger* class faster, although it was outgunned by the Royal Navy's *Queen Elizabeth* class of 1915. Nevertheless, the High Command thought them worthy of joining in the line of battle.

For 65 per cent of its length the ship was double-bottomed, and the voids were used for coal stowage. Originally, extendable torpedo nets were stowed along the sides. These were removed during 1916, as with most other capital ships, following the introduction of wire-cutting torpedoes.

The 305mm (12in) gun turrets traversed by electric power but the guns elevated hydraulically (maximum 13.5°, increased to 16° in 1916), firing 405.5kg (894lb) AP shells at a rate of up to three per minute and with a maximum range from 1916 of 20,400m (22,300yds). Their magazines held 720 shells. Incidentally, the normal barrel life of these guns was 200 rounds.

Engine power

The original plan was for triple-screw propulsion, with a central propeller driven by a diesel engine, to be used for cruising, and the two outer ones by steam turbines. Delays to the diesel engine brought a change to four-shaft, all-turbine drive. High pressure steam drove the outer shafts; low pressure

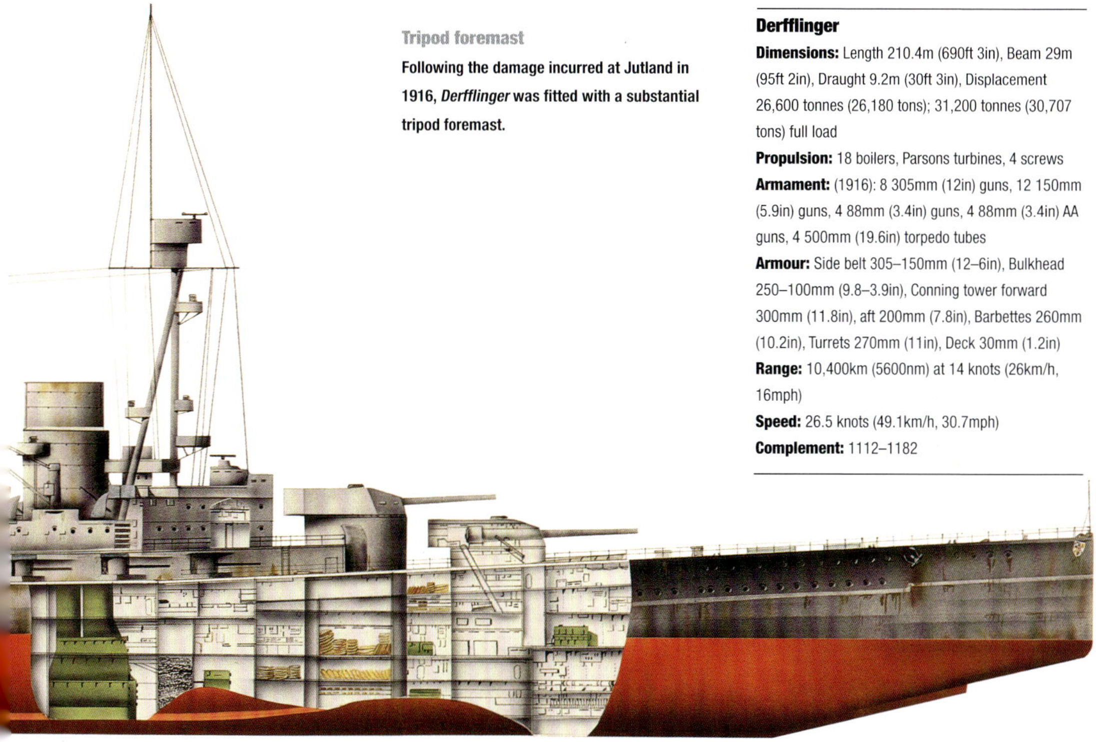

WORLD WAR I

Tripod foremast
Following the damage incurred at Jutland in 1916, *Derfflinger* was fitted with a substantial tripod foremast.

Derfflinger
Dimensions: Length 210.4m (690ft 3in), Beam 29m (95ft 2in), Draught 9.2m (30ft 3in), Displacement 26,600 tonnes (26,180 tons); 31,200 tonnes (30,707 tons) full load
Propulsion: 18 boilers, Parsons turbines, 4 screws
Armament: (1916): 8 305mm (12in) guns, 12 150mm (5.9in) guns, 4 88mm (3.4in) guns, 4 88mm (3.4in) AA guns, 4 500mm (19.6in) torpedo tubes
Armour: Side belt 305–150mm (12–6in), Bulkhead 250–100mm (9.8–3.9in), Conning tower forward 300mm (11.8in), aft 200mm (7.8in), Barbettes 260mm (10.2in), Turrets 270mm (11in), Deck 30mm (1.2in)
Range: 10,400km (5600nm) at 14 knots (26km/h, 16mph)
Speed: 26.5 knots (49.1km/h, 30.7mph)
Complement: 1112–1182

the inner two, which were geared turbines used for slower cruising. Steam came from a combination of 14 Admiralty double-ended coal boilers and eight double-ended oil boilers. Electric power was generated by two diesel engines driving two turbo-electric generators. Fuel capacity was 3500 tonnes (3400 tons) of coal and 1000 tonnes (984 tons) of oil.

This aerial view was probably taken in 1919 when *Derfflinger* was interned at Scapa Flow.

WORLD WAR I

Kongō (1913)

JAPAN

Intended to outclass the Royal Navy's *Invincible*, *Kongō* was also more than a match in firepower, if not speed, for the *Lion* class.

Built by Vickers at Barrow-in-Furness, in England, and completed on 6 April 1913, the designer was Sir George Thurston, the chief naval architect at Vickers. Three others of the same class were subsequently built in Japan. Britain attempted to get a loan of the *Kongō* class ships in 1915, but Japan rejected the idea.

Japanese gun intelligence
The original plan was for eight 305mm (12in) guns but secret information from Japanese agents revealed that the British Admiralty was planning 330mm (13.5in) for the *Lion* class battlecruisers, and so Japan opted for 356mm (14in).

Kongō's 356mm (14in) guns had the longest range of any at the time, up to 35.45km (22,028yds, 19.14nm). Its eight 533mm (21in) submerged torpedo tubes were the most ever fitted to a major warship.

Power came from 36 Yarrow coal-fired boilers with oil spray, driving two sets of Parsons direct-drive turbines.

Fast battleship
Between 1929 and 1931 *Kongō* and three sister ships were rearmed and reconstructed as fast battleships, although increased armour and anti-torpedo bulges actually slowed them. The funnels were reduced to two. Fire control systems improved, although this was an area where Japan lagged somewhat. Stroud & Barr had provided fire control tables for *Kongō* in 1913: these were probably used as the basis for tables supplied by Aichi Co. in 1933.

Another drastic modernization was made in 1935–37. The stern was extended and reshaped. Sixteen new Kampon oil-fired boilers and Brown-Curtis direct-drive turbines doubled the power output, with speed increased to 30.5 knots (56km/h). The secondary battery was revised, with 8 127mm (5in) dual-purpose guns, and revised again after 1942 to eight 152mm (6in), eight 127mm (5in) and 122 type 96 rapid-fire AA cannon.

New appearance
By 1937 the ship looked radically different. One massive funnel replaced the former two. The mainmast was removed and the foremast replaced by a multi-level pagoda structure topped by a massive rangefinder. A launching catapult for floatplanes, installed between the after turrets in the earlier refit, was enlarged. Only the

WORLD WAR I

A photograph of Kongo taken after her 1929–31 reconstruction.

Kongō

Dimensions: Length 222m (728ft 4in), Beam 31m (101ft 8in), Draught 9.7m (31ft 10in), Displacement 37,187 tonnes (36,600 tons)
Propulsion: 36 Yarrow water-tube boilers, 2 sets of Parsons direct-drive turbines, 4 screws
Armament: 8 356mm (14in) guns in twin turrets, 16 152mm (6in) guns, 8 76mm (3in) guns, 8 533mm (21in) torpedo tubes
Armour: Belt 279–203mm (11–8in), Deck 58-38mm (2.3–1.5in), Turrets 229mm (9in), Barbettes 254mm (10in)
Range: 19,000km (11,875mi, 10,000nm) at 14 knots (26km/h, 16.25mph)
Speed: 27.5 knots (50.9km/h, 31.6mph)
Complement: 1360

forecastle shape gave any indication of the ship of 1913.

The four *Kongō* class ships became Third Division of the First Fleet (main battle fleet). *Kongō* had a very active career in World War II until it was sunk by two torpedoes from US submarine Sealion on 21 November 1944.

Original profile

Kongō's original appearance. The elevated structure between the first two funnels is the compass platform, mounted clear of magnetic interference.

WORLD WAR I

Royal Oak (1914)

UNITED KINGDOM

Eight ships were originally to form the Revenge class, but only five were built. *Royal Oak* **was laid down at Devonport Naval Dockyard on 15 January 1914, launched on 29 April 1915 and completed in May 1916.**

Its cost was £2,468,269. The 381mm (15in) guns were the same as on the *Queen Elizabeth* class. Secondary armament was 14 152mm (6in) guns: the last time a main-deck level battery was included on a Royal Navy battleship, although they were set further back from the bows than on previous ships to minimize drenching by waves and spray. It had two 76mm (3in) AA guns and four 3-pounder guns. Four underwater 533mm (21in) torpedo tubes were fitted, with 20 torpedoes. These were removed by 1930, but in 1934–35 four torpedo tubes were fitted in the bow above the waterline, pointing abeam on each side.

The blister shape of the torpedo bulge is apparent. Total armour weight was 8250 tons (8382 tonnes).

Switch to oil

Mixed coal-oil firing of the boilers was in the original plan, as high speed was not a requirement, but when Admiral Fisher was reappointed as First Sea Lord in 1914, he demanded a change to oil only, giving an additional 2 knots (3.7km/h) of speed. With three boiler rooms next to one another, the flues were trunked into a single massive funnel. Searchlight positions were built up about the after casing of the funnel. By 1939 *Royal Oak* was the only ship in the class not to have a black-painted peaked cowl on the funnel-top. Beginning with HMS *Ramillies*, the class were the first battleships to be fitted with anti-torpedo bulges ('blisters'), adding 4m (13ft) to the beam, and in *Royal*

48

Oak the bulges were heightened in 1927, almost reaching the battery deck. As on other RN battleships, the original secondary armament was progressively reduced and additional AA guns fitted.

Modernizations

Between 1917 and 1923 all ships of the class had aircraft take-off platforms fitted to the superfiring turrets; and in 1934–35, *Royal Oak* had a catapult mounted on X turret, and an aircraft crane was placed on the port side of the mainmast. Other modernizations made at Devonport at this time were a redesigned bridge structure, improved wireless communications and improvements to AA defences, including the mounting of an eight-barrelled, two-pounder pom-pom on each side of the funnel, along with two quadruple-mount 18mm (0.7in) machine guns on each side of the conning tower. Up to 1938 the mainmast was of pole type, supporting a derrick, but when gunnery control instrumentation was installed at the crosstrees it was fitted with additional supports as a tripod. By this time, too, additions to the tower platforms and their combining into a single housing made tower and foremast into an integrated structure. Nevertheless, in 1939 the extent of modernization on board *Royal Oak* was quite limited. It still had the original engines, and its speed was no longer sufficient to maintain station with the fleet.

With the outbreak of war, it was moved to the Scapa Flow anchorage, where it was torpedoed in a bravura operation on the night of 14 October 1939 by *U-47*. A first torpedo hit the bows, followed by a second salvo, which scored three hits, exploding beneath the ship's bottom. It capsized and sank within 10 minutes, with the loss of 833 lives.

Main gun
The 381mm (15in) Mk I 42 cal gun was in use by the Royal Navy from 1915 to 1959 and proved an effective weapon in both world wars.

Royal Oak
Dimensions: Length 189m (620ft 6in), Beam (with bulge) 31.2m (102ft 2in), Draught 8.7m (28ft 6in, Displacement 29,110 tonnes (28,650 tons); 34,037 tonnes (33,500 tons) with bulges
Propulsion: 18 Yarrow boilers, Parsons geared turbines, 4 screws, 30,000kW (40,000shp)
Armament: 8 381mm (15in) Mk1 guns, 14 152mm (6in) guns, 8 102mm (4in) guns, 4 533mm (21in) torpedo tubes
Armour: Belt 330mm (13in), Turret faces 330mm (13in), Barbettes 254mm (10in), Bulkheads 152mm (6in), Deck 51–25mm (2–1in)
Range: 7400km (4200nm) at 10 knots (18.5km/h/11.5mph)
Speed: 23 knots (42.6km/h)
Complement: 997

WORLD WAR I

Texas (1914)

UNITED STATES

Texas (BB-35), the only remaining battleship to have fought in both world wars, survives as a museum ship at La Porte, Texas.

It was built at Newport News, Virginia, and commissioned on 12 March 1914. One of the two ships in the *New York* class, in general design it resembled other US ships of the era, with a level bow and a flush deck, with two funnels fitted closely between a pair of basket masts. An enclosed pilot house was built above the conning tower and attached to the foremast in 1917.

Turret design

One design proposed seven 305mm (12in) twin turrets but instead the ships were fitted, for the first time, with 356mm (14in) 45 calibre guns in five twin turrets, giving it the capacity to fire a broadside of 10 AP 6350kg (1400lb) shells to a range of 21km (13mi). These turrets, with inward-sloping sides, were known as 'turtlebacks'. Spiral grooves ('rifling') inside the barrels forced the projectile into rotating as it flew out, helping to keep it in a straight line. Turbine drive was in the original specification but two vertical triple-expansion engines were fitted, fired originally with 14 boilers. *Texas* and *New York* were the US Navy's last coal-fired ships.

Texas was the first US ship to be fitted with an experimental rangekeeper, which did little more than generate the angle of fire, but improvements in gunnery control were to follow.

Major modernization

A major modernization between 1925 and 1927 saw tripod masts installed, conversion to oil fuel with six new Bureau Express boilers and a single funnel, an aircraft catapult on the midships turret, with adjacent crane, and anti-torpedo blisters. Built-up range-finding and fire control stations on the foremast contributed to the change in appearance. Further changes and additions were relatively minor and sometimes temporary, such as the siting of AA guns on the superfiring turrets, although the upper mast section was removed in 1933–34. Two searchlights were fitted on the foremast, one above the other.

Quick-firing high-elevation 76mm (3in) tower-mounted AA guns, capable of 30 shots a minute, were tested on *Texas* in August 1916. 76mm (3in) guns were fitted on turrets 3 and 4 in 1921, and a torpedo defence platform was installed below the searchlight platform on the foremast in 1919–20, with a wood-plank aircraft launch platform on turret two in November–December 1918. Turret 3 also acquired a launch platform. The first flight off a US battleship was made by a Sopwith Pup from *Texas* on 10 March 1919.

World War II profile
USS *Texas* as it appeared in 1944.

WORLD WAR I

War action

In 1918, *Texas* crossed the Atlantic to join the British Grand Fleet in the 6th Battle Squadron. In World War II, it was employed as a convoy escort and also gave support to the shore landings of Operation Torch (1942) and D-Day (1944). Deployed to the Pacific in 1945, it fought in the Battle of Iwo Jima and gave support to the landings on Okinawa.

Texas was fitted with radar from 1938, initially an experimental bridge-mounted CXZ antenna, replaced by 1943 with SRa forward, and SRa and SC aft. In 1948 SK (aft) was also fitted.

USS *Texas* in the 1920s, prior to the 1926–27 refit.

Main tower

The tower abaft the funnel housed fire control radar Mk 3, 356mm gun director Mk 20 Mod 1, and 127mm gun director Mark 6 Mod 7.

USS Texas

Dimensions: Length 174.5m (572ft 7in), Beam 29m (95ft 3in), Draught 8.7m (28ft 5in), Displacement 27,433 tonnes (27,000 tons)
Propulsion: 14 boilers, 2 vertical triple-expansion engines, 2 screws, 20,954kW (28,100shp)
Armament: 10 356mm (14in) guns, 21 127mm (5in) guns, 4 533mm (21in) torpedo tubes
Armour: Belt 304–254mm (12–10in), Deck 76mm (3in), Turrets 356mm (14in), Barbettes 305mm (12in), Conning tower 305mm (12in)
Range: 14,816km (8,000nm) at 10 knots (18.5km/h, 11.5mph)
Speed: 21 knots (39km/h, 24.4mphk)
Complement: 1530

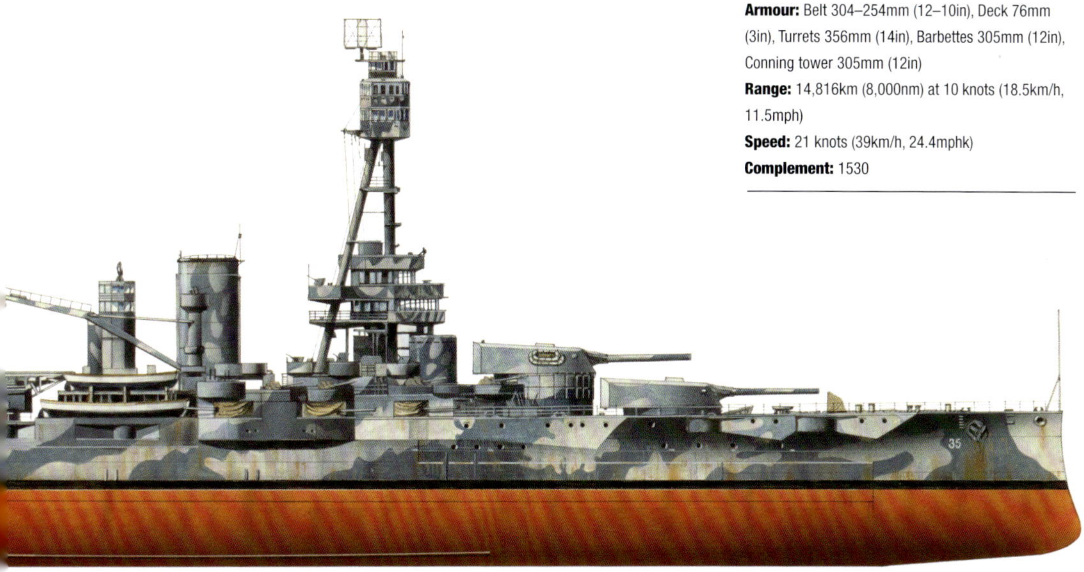

WORLD WAR I

Andrea Doria (1915)

ITALY

Most battleships of World War I that were not decommissioned underwent major changes of appearance and equipment in the 1920s.

1940 profile
Andrea Doria as it appeared from 1940.

Andrea Doria, designed by Vice-Admiral Giuseppe Valsecchi and built at La Spezia, retained its form as commissioned on 30 June 1916, until a virtual rebuilding in 1937–40. By then it was very old-fashioned, with a ram-type bow, minimal superstructure and two widely spaced funnels set close behind a pole foremast and tripod mainmast; two turrets fore and aft with superfiring and a central main turret that revolved to fire across a wide arc to port or starboard. Turrets A, X and the central one, Q, had triple mounts, giving the ship 13 305mm (12in) guns. The forecastle deck extended to the forward funnel; secondary 152mm (6in) guns were in casemates below, with the usual problems of wave action interrupting fire. All the guns were built by Vickers and Armstrong Whitworth in England (Italy was an ally in 1914–18).

New look ship
The ship that emerged from the Trieste yard was 11.3m (37ft) longer and 6100 tonnes (6024 tons) heavier, with a redesigned bow and stern and the forecastle deck extended back to the mainmast. Most of the weight increase was accounted for by heavier armour, notably on the deck. The danger of plunging fire with AP shells was by now well understood. Eight oil-fired boilers replaced 20 mixed-fuel burners, and two new Belluzzo reduction-gear turbines delivered a speed of 28 knots (48km/h, 30mph) and tripled the original power output of 22,371kW (30,000shp). Q turret, the casemate guns, and the three 450mm (17.7in) torpedo tubes were all removed, the two funnels were set close together, the superstructure was built up to hold the control posts of a modern warship, and an aft control tower replaced the mainmast, with a short topmast. Twelve 135mm (5.5in) guns mounted in triple turrets amidships, and a substantial anti-aircraft battery, were also installed. Compared to the British flagship in the Mediterranean, HMS *Queen Elizabeth*, *Andrea Doria* was inferior in gun power but considerably faster.

Diplomatic duties
Andrea Doria saw no action in World War I and carried out mainly diplomatic duties in the 1920s, apart from shelling rebel positions at Fiume (Rijeka) in 1920. Convoy escort and deterrent movements against British warships were its tasks in World War II. Interned after Italy's surrender, it was reinstated after 1945 and was flagship of the *Marina Militare* in 1949–50 and 1951–53. It was scrapped in 1956.

WORLD WAR I

Andria Doria (post-1940)

Dimensions: Length 186.9m (613ft 2in), Beam 29.2m (95ft 3in), Draught 8.6m (28ft 3in), Displacement 29,100 tonnes (28,634 tons); 29,400 tonnes (28,929 tons) full load
Propulsion: 8 Yarrow superheated boilers, 2 Belluzzo reduction gear turbines, twin screws, 67,113kW (90,000shp)
Armament: 13 305mm (12 guns), 2 triple and 2 twin turrets, 12 135mm (5.5in) guns, 10 90mm (3.5in) 12 37mm (1.45in), 15 20mm (.78in) AA guns
Armour: Belt 250–130mm (10–5.1in), Barbettes 280–130mm (11–5.1in), Deck 166–135mm (6.5-5.3in), Conning tower 250mm (10in)
Range: 12,408km (6700nm)
Speed: 28 knots (48km/h, 31mph)
Complement: 1300

Andrea Doria in early fitting-out stage, with funnels installed.

WORLD WAR I

Queen Elizabeth (1915)

UNITED KINGDOM

This was the British Royal Navy's first oil-fired battleship, and the first with 381mm (15in) guns.

Quickly termed a 'super-dreadnought', this was the leader of a class of five fast battleships whose design was driven by the demands of impending warfare, although none of them attained the design speed of 25 knots (46.3km/h, 28.8mph), due to excessive loading and a draught greater than intended.

Tonnage for offence
In battleship construction, the British were more inclined to devote tonnage to offensive purposes rather than to defence. By contrast, the Germans gave more attention to armour and internal compartmentalization, which proved to be a better policy, particularly when linked to a closer focus on important details, such as gun management and firing procedures. The British 381mm (15in) gun was developed with remarkable speed in response to Germany's reported adoption of the same size in SMS *Bayern*. Modelling and testing were telescoped, and the 343mm (13.5in) gun of immediate predecessors, such as HMS *Iron Duke*, were used as development models.

It proved to be an excellent weapon. Weighing 98.5 tonnes (97 tons), it fired AP shells of 871kg (1920lb) with a muzzle velocity of 750m/sec (2450ft/sec) and a range of 30,680m (33,550yds). For the Mark I, elevation was limited to 20°, although this was later raised to 30°. They were set in superfiring turrets, two forward and two aft.

1940 profile
Queen Elizabeth in post-1940 form.

HMS Queen Elizabeth
Dimensions: Length 195.34m (640ft 11in), Beam 27.6m (90ft 6in), Draught 9.1m (30ft), Displacement 27,940 tonnes (27,500 tons); 33,548 tonnes (33,020 tons) full load
Propulsion: 24 Babcock & Wilcox boilers, 4 Parsons geared turbines, developing 55,927kW (75,000shp), 4 screws
Armament: 8 381mm (15in) guns, 16 152mm (6in) guns, 2 76mm (3in) AA guns, 4 533mm (21in) torpedo tubes
Armour: Belt 330–102mm (13–4in), Bulkheads 152–102mm (6–4in), Barbettes 254–102mm (10-4in), Turrets 330–127mm (13–5in), Deck upper 45–32mm (1.7–1.2in), Deck lower 25mm (1in) with 76mm (3in) over steering gear
Range: 13,840km (7,500nm) at 12.5 knots (23.2km/h, 14.5mph)
Speed: 25 knots (46.3km/h, 28.9mph)
Complement: 951

In its original form, the ship had a more substantial superstructure than its predecessors, reflecting the need to mount a more complex range of range-finding and fire control equipment. Two funnels were set between a tripod foremast and a taller mainmast (*Iron Duke* had only a foremast but this was unusual).

Two fire control directors were installed from the start, on the mainmast, with a 2.7m (9ft) rangefinder, and a larger one (4.6m, 15ft) above the conning tower.

In addition, each turret was equipped with its own 4.6m (15ft) rangefinder. A torpedo control director was placed aft of the conning tower. An electrically-powered Dumaresq MkIV fire-control table was installed in the conning tower.

Modifications

In early 1919, aircraft flying-off platforms were fitted to B and X turrets and a series of major modifications went on in the 1920s, followed in 1937–40 by new engines and boilers. The original boilers and direct-drive turbines were replaced by eight 3-drum, small-tube superheated boilers with air pre-heaters and a working pressure of 28kg.sq cm(400psi), and geared turbines. The saving of usable space amounted to one-third, and the new installation weighed half of its predecessor.

Queen Elizabeth was seriously damaged by Italian frogmen at Alexandria on 19 December 1941. Out of action until June 1943, it then served mostly in the Indian Ocean. It was scrapped in 1948.

Underwater defences

The ship's underwater defences were substantial, but the Italians' explosives blew two holes in the double bottom. As a result, the ship was out of combat action for 18 months.

WORLD WAR I

Canada/Almirante Latorre (1915)

UNITED KINGDOM/CHILE

Ordered by the Chilean government from the Tyneside commercial shipyard of Armstrong Whitworth (specialists in warship building) in 1911 and launched as *Almirante Latorre* on 27 November 1913, the ship was still fitting out when Britain declared war on Germany on 4 August 1914.

By agreement between the two governments, it was taken over by the Royal Navy and renamed HMS *Canada*. It was commissioned on 15 August 1915.

One-off battleship

Canada was a one-off, the only British battleship of the time with 355.6mm (14in) guns. While conforming to the requirements of the Chilean Navy, these were essentially to obtain a modern British-style battleship, and there were broad similarities with the *Iron Duke* class of 1914, with a raised forecastle above a long flush deck on which the 152mm (6in) guns were mounted on barbettes. Proportions were slightly different, with *Canada* having a shorter forecastle and a longer quarterdeck, and with a draught 1.2m (3ft 6in) less, perhaps because coastal bombardment was intended as one of the ship's main functions. A tripod foremast was set close to the main superstructure, and the funnels were heightened to improve draught in the boilers. Following the handover, a number of modifications were made, including removal of the bridge and the placing of two open platforms on the foremast, and a boat derrick between the funnels.

Its armour was comparatively light, comparable to that of a battlecruiser rather than a battleship, again perhaps anticipating fighting conditions in the eastern Pacific, but this did not deter the Admiralty from sending the ship to fight at Jutland.

The machinery comprised 21 mixed-firing Yarrow boilers, supplying steam to two Brown-Curtis high pressure and two Parsons direct low-pressure turbines, giving 27,591kW (37,000shp) and maximum speed of 22.75 knots (42.1km/h, 26.2mph). Bunker capacity was 3300 tonnes (3254 tons) of coal and 520 tonnes

---

HMS Canada/Almirante Latorre
Dimensions: Length 191m (625ft), Beam 28.2m (92ft 6in), Draught 10m (33ft), Displacement 25,400 tonnes (25,000 tons); 32,514 tonnes (32,000 tons) full load
Propulsion: 21 Yarrow boilers, 2 Brown-Curtis HP and 2 Parsons LP turbines, 4 screws, 27,591kW (37,000shp)
Armament: 10 356mm (14in) guns in twin turrets, 16 152mm (6in) guns, two 76mm (3in) and 4 3-pounder guns, 4 533mm (21in) torpedo tubes
Armour: Belt 230mm (9in), Barbettes and turrets 254mm (10in), Conning tower 280mm (11in)
Range: 8153km (4350nm)
Speed: 22.75 knots (42.1km/h, 26.2mph)
Complement: 834

WORLD WAR I

(509 tons) of oil, allowing an operating range of 8153km (5066mi, 4350nm) at 10 knots.

Elswick built the 356mm (14in) guns, including four spare barrels that were never fitted. With a maximum elevation of 20°, they could fire 719kg (1516lb) AP shells a distance of 22,300m (24,400yds), at a muzzle velocity of 764m per second (1507 ft/s). This was the last Royal Navy battleship to have a fifth turret set between deck structures. This had only a limited angle of fire because of potential blast effects against the deckhouses, but gave *Canada* a formidable 10-gun broadside. The 152.4mm (6in) guns fired 45.36kg (100lb) shells to a range of 16,000m (18,000yds).

Regains original name

In 1921, the ship was finally purchased by Chile and regained its original name. In 1929–31, a modernization was carried out at Plymouth. With new oil-fired boilers and turbines, it also acquired anti-torpedo bulges, a higher mainmast (18.3m/60ft) and AA guns mounted on the superstructure. Little used in the 1930s, it was obsolescent by 1941, when the USA sought to buy it to tide over its Pearl Harbor losses. It was finally decommissioned in 1958 and scrapped in 1959.

Almirante Latorre, c. 1948, its appearance scarcely changed since 1931.

Dual service
HMS *Canada* as it appeared just before being handed back to Chile.

WORLD WAR I
Fusō (1915)

JAPAN

While the battleship-building race between the two main North Sea powers was intensely public, a quieter but equally dramatic race was shaping up in the Pacific Ocean, between Japan and the United States.

Fusō, the first Japanese 'super-dreadnought' was built to rival USS *Texas* and *New York*. Designed wholly in Japan, its builders drew heavily on British and European practice. Laid down in 1912, *Fusō* was not commissioned until 1917, and played no more than a patrolling role in World War I.

British profile
Its original profile was very similar that of a British battlecruiser, with a low conning tower and navigation deck, a tripod foremast, two widely spaced funnels and a short tripod mainmast. Twelve 356mm (14in) guns were mounted, with superfiring twin turrets fore and aft, and two central turrets on slightly different levels. All were on the centre-line.

Steam was raised in 24 Miyahara mixed-fuel water-tube boilers, to power twin sets of Brown-Curtis direct-drive turbines. 4100 tonnes (4000 tons) of coal and 1016 tonnes (1000 tons) of fuel oil were carried, giving a potential range of 15,000km (8000nm).

Two-stage rebuild
Fusō retained its original form and fittings until a two-stage rebuild, in 1930–33 and 1935–37, dramatically altered its appearance and fighting capacity. Unusually, the rebuild reduced its displacement by 1290 tonnes (1270 tons). New 356mm (14in) guns were fitted, with the turrets modified to reach 43° elevation and increasing range to 33,000m (30,000yds). Six Kampon superheated boilers allowed the elimination of one funnel, and four Kampon steam turbines provided the drive. The stern was extended by 7.62m (25ft). The most striking feature was the multi-level 'pagoda' assembled around the now invisible foremast, an intentionally intimidating seaborne war tower. Torpedo bulges and additional armour were also added, with deck armour almost doubled in thickness. A catapult and hangar space for three seaplanes were provided. Oil bunkerage was 5200 tonnes (5000

Early profile
Fusō as originally built, 1915.

tons), enough to keep the ship at sea for 21,900km (13,600mi, 11,800nm).

Ability to search

The impressive tower of *Fusō* rather exaggerated the Japanese ability to search and detect other ships. Night fighting had been closely planned by the Imperial Navy (IJN), helped by the high quality of the optics produced by Nippon Kōgaku KK (Japan Opticazl Co.), which included the type 88 Model 1 (1932) with 120mm (4.7in) lenses and enhanced light-gathering capacity for night vision. The IJN also developed a parachute-suspended star-shell to illuminate a target without displaying the source direction. However, the superiority of US radar annihilated any advantage.

Fusō (1937)

Dimensions: Length 212.75mm (698ft), Beam 10.08m (33ft 1in), Draught 9.69m (31ft 9in), Displacement 39,773 tonnes (39,145 tons) full load
Propulsion: 6 Kampon boilers, 4 Kampon geared steam turbines, 4 screws, 55,927kW (75,000shp)
Armament: 12 356mm (14in) guns, 14 152mm (6in) guns (1938); 8 127mm (5in) guns, 16 132mm (5.2in) AA guns in quadruple mountings
Armour: Side belt 305–102mm (12–4in), Main turrets 297–114mm (11.7–4.5in), Casemate 152mm (6in), Deck 132–51mm (5.2–2in)
Aircraft 3 floatplanes
Range: 21,900km (13,600nm) at 16 knots (29.6km/h, 18.5mph)
Speed: 27.74 knots (51.4km/h, 32.1mph)
Complement: 1396

Late profile
Fusō's appearance post-1937.

WORLD WAR I

Nevada (1916)

UNITED STATES

Nevada (BB-36) was designed in 1910 and built at the Fore River shipyard. This ship exemplified a development in armour plating that became known as the 'all or nothing' approach.

Reconstruction
USS *Nevada* after reconstruction, May 1943.

While US warship designers were sometimes said to wait for other navies to test and develop useful new features, this was by no means always the case, as *South Carolina* showed. *Nevada* and its class sister *Oklahoma* demonstrated a new approach to the application of armour plating, with heavy protection for engines and magazines, and light protection for less vulnerable parts. This became known as the 'all or nothing' approach, doing away with the gradual thinning out of the armour, but providing a really solid armoured box to protect the engines and magazines, with 344mm (13.5in) transverse bulkeads fore and aft.

US problem-solving
This large compartment also provided enough reserve buoyancy to keep the ship afloat, as well as functioning, in the event of flooding, in the less protected parts. It was a very intelligent solution to the constant problem of balancing weight, machinery and armament, taking account of the realities of long-range fire and heavy armour-piercing shells. Perhaps because it was a US concept, it took the British Navy several years, and the Battle of Jutland, to follow suit in battleship design.

The 356mm (14in) guns were now mounted in two forward and two aft turrets, the lower two having triple mounts, with twin mounts in the superfiring ones. Another notable design change was the abandoning of the flush deck for a long foredeck and shorter afterdeck one level lower. This made it possible for the original casemate-mounted guns to be raised in a 1927–29 refit. Both ships were oil-fired from the start, but the original boilers on *Nevada*, were replaced by six Bureau Express models, and Curtis geared steam turbines taken from USS *North Dakota*.

Damage at Pearl Harbor
Basket masts were exchanged for tripods in the first refit, the torpedo tubes originally fitted were removed, anti-torpedo bulges were added to the sides, and two aircraft catapults were fitted, one on X turret and one on the quarterdeck. *Nevada* was damaged at Pearl Harbor and subsequently rebuilt. The superstructure was redesigned to provide more mounts for AA guns, the bridge was extended backwards to meet the funnel, which was given a unique backwards-tilted smoke tube,

USS Nevada

Dimensions: Length 177.8m (583ft 4in), Beam 29m (95ft 3in), Draught 8.7m (28ft 5in), Displacement 27,941 tonnes (27,500 tons); 29,364 tonnes (28,900 tons) full load
Propulsion: 12 Yarrow boilers, 2 Curtis turbines plus 2 geared cruising turbines, 2 screws, 19,761kW (26,500shp)
Armament: 10 356mm (14in) guns, 21 127mm (5in) guns, 2 76mm (3in) AA guns, 2 533mm (21in) torpedo tubes
Armour: Belt 343–203mm (13.5–8in), Deck upper 76mm (3in), Deck lower 38mm (1.5in), Funnel uptakes 305mm (12in), Barbettes 343mm (13.5in), Turrets 457–127mm (18–5in)
Range: 18,520km (10,000nm) at 10 knots (18.5km/h, 11.5mph)
Speed: 20.5 knots (38km/h, 23.5mph)
Complement: 864

USS *Nevada* at anchor, Panama Bay, mid-1920s.

the catapult on X turret was removed and the 21 127mm (5in) guns taken out, but AA firepower was provided by 48 40mm and 27 20mm AA guns. SRa and SK radar antennas were also fitted. *Nevada*'s armour protection withstood torpedo and bomb hits at Pearl Harbor and a kamikaze attack in March 1945. Its last uses were as a target in the Bikini Atoll A-bomb tests of 1946, when it remained afloat, and was finally sunk in July 1948 in aerial torpedo practice.

WORLD WAR I

Bayern (1916)

GERMANY

The 'super-dreadnoughts' *Bayern* and *Baden* – the only two completed of a planned class of four – have always been compared with the coeval British *Queen Elizabeth* class, of similar tonnage and dimensions, and also the same main armament of eight 381mm (15in) guns.

The ships never fought each other, *Bayern*'s completion coming too late for it to participate in the Battle of Jutland. *Bayern* was shorter than *Queen Elizabeth* but broader-beamed, indicative of the fact that armour protection and internal subdivision were more substantial and comprehensive on the German ship.

Construction

Built at the Howaldtswerke, Kiel and commissioned in July 1916, *Bayern* was constructed of riveted steel plates attached to longitudinal and transverse bulkheads, with a cellular double bottom along almost 90 per cent of the hull. Internally it was divided into 17 watertight compartments. Alongside the boiler and engine rooms there was an inner torpedo bulkhead, flanked on the inboard side by coal bunkers. The superstructure was low, with a rangefinder tower and navigation bridge just one level above the forward superfiring turret, which might have caused blast problems

Four turrets

Bayern's four turrets and their guns remain on the seabed in Scapa Flow, Orkney, having dropped out when the overturned ship was partially raised in 1934.

SMS Bayern

Dimensions: Length 180m (590ft 7in), Beam 30m (98ft 5in), Draught 9.4m (30ft 10in), Displacement 32,200 tonnes (31,700 tons)

Propulsion: 14 marine boilers, 3 Schichau geared turbines, developing 38,776kW (52,000shp), 3 screws

Armament: 8 381mm (15in) guns, 16 152mm (6in) guns, 2 88mm (3.5in) guns, 5 600mm (24in) torpedo tubes

Armour: Belt 350–170mm (13.8–6.7in), Bulkheads 200–170mm (7.9–6.7in), Deck 150–30mm (5.9–1.2in), Barbettes 350mm (13.8in), Turrets 350–250mm (13.8–9.8in) Conning tower: forward 350mm (13.8in); aft 170mm (6.7in)

Range: 14,500km (7800nm) at 10 knots (18.5km/h, 11.5mph)

Speed: 22.25 knots (41.2km/h, 25.6mph)

Complement: 1171

in combat. A tripod foremast with a high topmast and a shorter pole mainmast were separated by two funnels of equal height but of smaller diameter in the aft one, which had boat derricks mounted alongside. The main guns were mounted in twin superfiring turrets fore and aft, the turrets projecting backwards on the barbettes, and no two turrets at the same level. The forecastle deck extended aft as far as X turret, above a long casemate with eight 152mm (6in) guns on each side.

Boiler power
Bayern's machinery comprised eleven coal-fired and three oil-fired Schulz-Thornycroft marine boilers (one of the numerous variants of the three-drum boiler), along with three sets of Schichau-Parsons steam turbines driving three shafts, developing 38,776kW (52,000shp). A diesel engine was also planned but a suitable engine had not been perfected by the time *Bayern* was fitting out. Eight small diesel generators were installed to provide electric power (2400kW capacity). Fuel bunkerage was 3400 tonnes (3300 tons) of coal and 620 tonnes (610 tons) of fuel oil.

There were few opportunities to make modifications before the war ended, but the main gun elevation was raised to 20°, extending maximum range to 23,200m (25,372yds), and four of the original five torpedo tubes were removed. From 26 November 1918, *Bayern* was interned at Scapa Flow, and scuttled by its crew on 21 June 1919.

Searchlights
Bayern was well-equipped with searchlights attached to the tripod supports, and davit-type cranes to raise and lower them.

WORLD WAR I

Provence (1916)

FRANCE

Built at the Arsénal de Lorient and its keel laid on 1 August 1912, *Provence* was first of the *Bretagne* class of three ships to be commissioned, on 1 March 1916.

The relatively slow pace of completion and smaller size of French battleships was caused at least in part by the country's lack of large fitting-out basins and dry docks to hold ships of increasing size. It was 1927 before the large Vauban basins at Toulon were completed.

Courbet class

The *Bretagne* class shared the hull type and general dimensions of the *Courbet* class. The number of turrets was reduced from six to five, to accommodate its heavier 340mm (13.4in) guns. A centre turret, firing to port or starboard, was placed between the two funnels, and all heavy guns were placed on the centre-line. Two pole masts of equal height were fitted.

Scuttled by its crew, *Provence* lies half-submerged at Toulon, 27 November 1942.

Early profile

Provence's appearance before 1935.

64

A substantial secondary armament was fitted at weather deck level. The ship was also equipped to carry and lay 28 mines. Armour protection was somewhat reduced compared with *Courbet*, to compensate for the greater weight of the guns.

Provence had 18 Guyot du Temple coal boilers, powering two sets of Parsons turbines. Fuel capacity was 2680 tonnes (2640 tons) of coal and 300 tonnes (295 tons) of oil, giving an operating range of 8500km (4600nm).

Inadequate range

The range of the 340mm (13.4in) guns was found to be inadequate and in 1919–20 the elevation was altered, but only to 18°, extending the range from 20,000m to 25,000m (21,870–27,340yds). New rangefinders and a fire-control director system were installed. In a 1925–27 refit, nine of the original boilers were replaced by oil burners.

The Model 1912M 340mm (13.4in) gun was used to rearm *Provence* in 1932–35 during a major refit at the Brest naval yard. The superstructure was rebuilt, a massive tripod foremast was fitted, modern fire control was installed and the central side armour was strengthened. AA defences were increased, with eight 75mm guns replacing older 138mm guns in the casemates, and 12 13mm machine guns mounted on the superstructure. The underwater torpedo tubes were removed.

War service

Provence served with the Mediterranean Fleet in 1914–18, participating in the blockade of the Adriatic, which bottled up the Austro-Hungarian fleet, although its activity was heavily limited by inability to obtain coal for the boilers. In Operation Catapult at Mers-el-Kebir on 3 July 1940, *Provence* was the

Provence

Dimensions: Length 166m (544ft 7in), Beam 26.8m (88ft 3in), Draught 8.9m (29ft 2in), Displacement: 23,936 tonnes (23,558 tons)
Propulsion: 18 Guyot du Temple boilers, 2 sets of Parsons direct-drive turbines, 4 screws, 22,000kW (29,000shp)
Armament: 10 340mm (13.4in) guns, 22 138mm (5.4in) guns in casemates, 4 47mm (1.8in) guns, four 450mm (17.7in) torpedo tubes
Armour: Belt 270–160mm (10.6–6.25in), Barbettes 248mm (9.75in), Deck 77–40mm (3.2–1.6in), Conning tower 340mm (13.4in)
Range: 8500km (4600nm) at 10 knots (18.5km/h, 11.5mph)
Speed: 20 knots (37km/h, 23mph)
Complement: 1174

first ship to return fire on the British attackers. After the ship was scuttled at Toulon in November 1942, the Germans removed the two aft turrets and set them up as coastal defence. In August 1944, they exchanged fire with *Provence's* sister battleship *Lorraine* and USS *Nevada*.

WORLD WAR I

Royal Sovereign (1916)

UNITED KINGDOM

Completed at Portsmouth Naval Dockyard in May 1916, *Royal Sovereign* just missed the Battle of Jutland, where its eight 381mm (15in) guns and up-to-date fire control equipment might have been useful to the Royal Navy.

Although the design period was pre-war, the ship, like others in the R-class (see HMS *Royal Oak*), can be seen to be taking on a more modern appearance, helped by the single funnel and the more substantial tower that now held control stations as well as the navigating deck. The illustration appears to show how it looked after a refit in 1927–29 when, among other things, new rangefinders and eight searchlights were fitted, and the tipped funnel cap was added. As with all new battleship classes, even in wealthy countries, the ultimate design was a compromise between what the designers and the seaman wanted, what the government was willing to spend and how the Admiralty had to divide its budget among a host of competing needs and demands. The R-class emerged as a scaled-down version of the *Queen Elizabeth* class – smaller, slower, but with good protection and endurance.

Propulsion

Motive power had originally been planned as coal-fired supplemented by oil, but Admiral Fisher, back as First Sea Lord, demanded full oil, and 18 Babcock & Wilcox boilers, working to a pressure of 235psi, were installed to drive two sets of Parsons direct-drive steam turbines. The turbines were in three separate compartments; the outer (high pressure) set drove the outer shafts and the low-pressure inner pair in the central room drove the inner two propellers at cruising speeds.

Fire control

Unlike earlier ships, *Royal Sovereign* was equipped with fire control systems from the start. A forward one was placed above the conning tower and the aft one was on the mainmast (both masts were tripods) with 4.6m (15ft) rangefinders. In the event of both being knocked out, there was a further control point in X turret. The torpedo control director was also placed aft, with a 2.7m (9ft) rangefinder. In the main control room was a Mark IV Dreyer table to

Late profile
The ship in post-1937 form. The oblong Carley floats, an American invention, were introduced on RN ships during the 1930s.

display the data collected, linked to an analogue computer. Data was also fed from Barr & Stroud Mark I telemeters in the turrets. Mark VII dumaresqs, for use in revolving control towers, were installed.

Flying platforms

In 1918, flying-off platforms were installed on B and X turrets, although the planes, Sopwith Pups and Camels, could only be used close to land. In its 1933–36 refit *Royal Sovereign* had a catapult and crane placed on the quarterdeck and could launch and recover floatplanes such as the Fairey III, but these were removed in 1937. A full refit intended for 1939 had to be cancelled by the outbreak of war.

In May 1944, the ship was secretly transferred to the Soviet Navy as *Archangelsk* and used for escorting convoys to Murmansk. Returned to Royal Navy in February

HMS *Royal Sovereign* c. 1936, perhaps at Devonport.

HMS Royal Sovereign
Dimensions: Length 190.3m (624ft 3in), Beam 27m (88ft 6in), Draught 8.7m (28ft 6in, Displacement 28,449 tonnes (28,000 tons); 31,497 tonnes (31,000 tons) full load
Propulsion: 18 oil-fired Babcock & Wilcox boilers, 2 sets Parsons direct-drive turbines, 4 screws, 29,828kW (40,000shp)
Armament: 8 381mm (15in) guns in twin turrets, 14 152mm (6in) guns, 2 76mm (3in) AA guns, 4 3-pounder guns, 4 533mm (21in) torpedo tubes
Armour: Belt 330–25mm (13–1in), Bulkheads 152–102mm (6–4in), Barbettes 254–102mm (10-4in), Turret faces 330mm (13in), Conning tower 280mm (11in)
Range: 7000nmi (13,000km; 8,100mi)
Speed: 23 knots (42.6km/h, 26.5mph)
Complement: 997

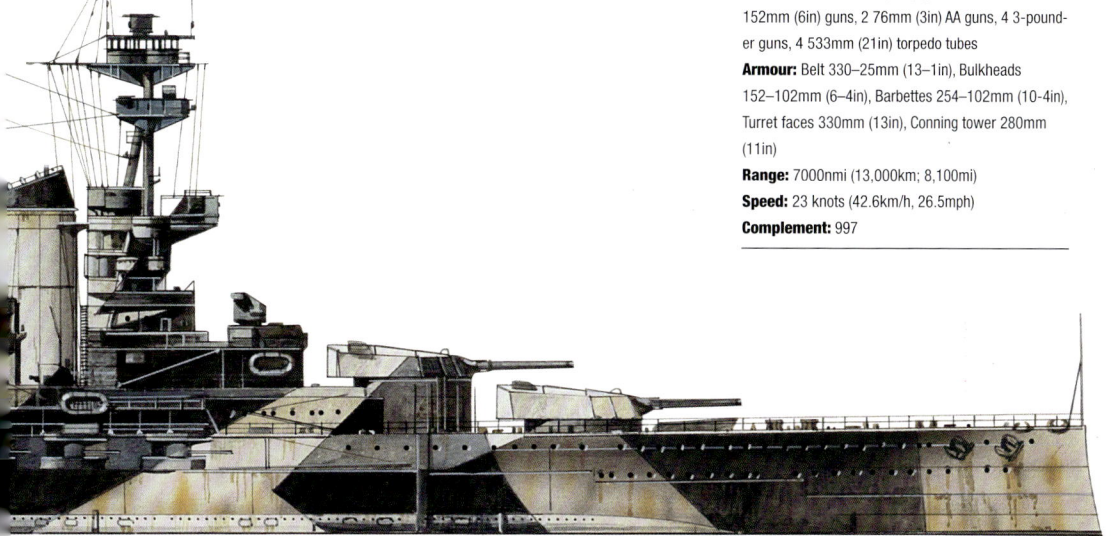

WORLD WAR I

1949, it was found to be in poor condition and was sent for scrapping that year.

Torpedo bulges

Torpedo bulges had first been fitted by the Royal Navy to shallow-draught big-gun coastal monitors operating in U-boat infested waters off the coast of Flanders in 1917. Their development, and perhaps the concept, is credited to Eustace Tennyson D'Eyncourt, Director of Naval Construction from 1912 to 1924. Flattening out fore and aft, they bulged outwards from the sides, intended to keep the centre of a detonation as far away as possible from the hull plating and to deflect the destructive energy upwards and downwards.

Bulge

Internally divided by a longitudinal watertight partition, the outer section of the bulge was empty and subdivided into watertight cells; the inner section was filled with water admitted through openings in the bottom. In normal loading, the bulge's top was just above the waterline. Speed was marginally reduced as a result (0.3 knots) but the staying power (and survival chances) of a ship in action were extended, and the concept was soon copied by other navies.

The name-ship of her class, HMS *Royal Sovereign* c. 1918-21. The upper edge of the torpedo bulge can be seen just at the waterline.

WORLD WAR I

Courageous (1917)

UNITED KINGDOM

Admiral Lord Fisher's name appears often in the Royal Navy's story between 1905 and 1918. The *Courageous* class battlecruiser was one of his less successful brainchildren, although his successors managed to adapt it to a different role.

Beam and length
The ship's beam/length ratio was not ideal for a flight deck (compared to *Ise*, for example) but its length certainly helped.

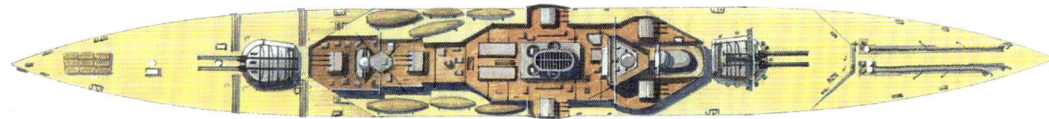

Major warships were generally intended to be able to fulfil a variety of roles, but the three *Courageous* class ships had the specific purpose of supporting a proposed Russian invasion of Germany's Baltic coast in 1917.

Designed for Baltic
They would land troops and carry 381mm (15in) guns to fight off German warships (when they were planned, no German battleship had 381mm guns). There were two twin turrets, fore and aft. In addition, 18 102mm (4in) guns were mounted in triple turrets, for the first time in an RN ship. An unusually long deckhouse allowed for troop accommodation and storage of materials. A maximum draught of 7.3m (24ft) was demanded of the designers, to navigate inshore Baltic waters: this was met, although the desired speed of 32 knots was not quite attained. Fisher's designation of the class was 'large light cruiser': this was a political term as government policy at the time prohibited any new ship larger than a light cruiser.

Armour protection was indeed notably light (see specification). Most of the armour was high-tensile steel intended to make HE shells detonate on impact: however, sea battles would increasingly be fought by AP shells in parabolic 'plunging fire', which *Courageous*'s deck would not withstand.

Krupp cemented armour was used on the barbettes and conning tower. This was the first Royal Navy ship to have Yarrow small-tube boilers and Parsons geared turbines. The four

WORLD WAR I

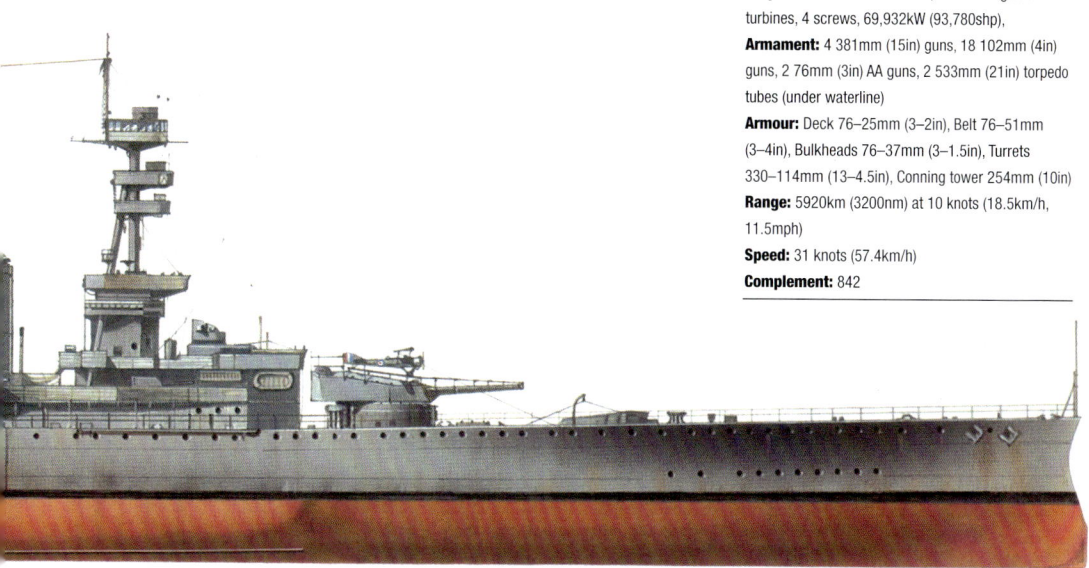

HMS Courageous (as battlecruiser)
Dimensions: Length 239.6m (786ft), Beam 24.7m (81ft), Draught 6.8m (22ft 3in), Displacement 17,527 tonnes (19,320 tons); 20,829 tonnes (22,960 tons) full load
Propulsion: 18 Yarrow boilers, 4 Parsons geared turbines, 4 screws, 69,932kW (93,780shp),
Armament: 4 381mm (15in) guns, 18 102mm (4in) guns, 2 76mm (3in) AA guns, 2 533mm (21in) torpedo tubes (under waterline)
Armour: Deck 76–25mm (3–2in), Belt 76–51mm (3–4in), Bulkheads 76–37mm (3–1.5in), Turrets 330–114mm (13–4.5in), Conning tower 254mm (10in)
Range: 5920km (3200nm) at 10 knots (18.5km/h, 11.5mph)
Speed: 31 knots (57.4km/h)
Complement: 842

turbines, each driving a propeller shaft, were in two rooms.

Eventual action

None of the class ever entered the Baltic Sea. *Courageous* was flagship of the First Cruiser Squadron in 1917, and took part in the inconclusive cruiser action in the Heligoland Bight on 17 November 1917. *Courageous* fired 92 rounds of 381mm (15in) and 180 102mm (4in) shells, but with only one hit, causing minor damage to the German cruiser *Pillau*. If three per cent of their rounds hit or near-missed an enemy, a ship was felt to have done well – the wastage was colossal, but effective fire control was still in its infancy. In 1917 *Courageous* was also fitted for minelaying, where speed was important, but was never used for the purpose. In 1918 flying-off platforms were installed on the turrets for Sopwith Pups and Strutters, which could take to the air at low speeds.

Between 1921 and 1929 the three ships were converted into aircraft carriers, for which their length and speed made them suitable. *Courageous* was sunk by torpedoes from *U-29* on 17 September 1939.

Light cruiser?

Fisher's definition of 'large light cruiser' raised a few eyebrows: light cruisers did not carry 381mm (15in) guns.

71

WORLD WAR I

Ise (1917)

JAPAN

Built by Kawasaki at Kobe, and commissioned on 1 December 1917, *Ise's* first incarnation was as a battleship armed with 12 356mm (14in) guns in twin turrets, a direct development of the preceding *Fusō* class.

Four of the turrets pointed aft in two superfiring sets separated by the mainmast and aft deckhouse with its control positions. Its first rebuild was at Kure Arsenal in 1927–28 when the pagoda-type superstructure that was to typify Japanese battleships was added. In 1935–37, at Kure, the original 24 coal/oil boilers were replaced by eight Kampon oil boilers and new Kampon turbines. Anti-torpedo bulges were fitted and its six torpedo tubes were removed.

Hybrid battleship/carrier
Although engaged in convoy escort and search-and-attack operations, *Ise* saw no combat action in 1942–43. After Japan's disastrous losses at the Battle of Midway the decision was made to convert the ship into a hybrid battleship/carrier to support

Ise, sunk off Kure, Japan, October 1945. Note the carrier *Amagi* also sunk in distance.

the depleted carrier fleet. Between February–August 1943 it was remodelled, retaining the front end and guns of a battleship but with the two aft turrets removed along with their barbettes (but not the magazines that now stored AA ammunition). A 70m (76.5yd) flight deck was built

Late profile
Ise as she appeared in October 1943.

WORLD WAR I

between the reduced mainmast and the stern, 6m (19ft) above the now-reinforced main deck and narrowing from 29m (31.7yd) to 13m (14.2yd) over the stern. Two Kure Type No.2 Model 5 trainable catapults were mounted, and a hydraulic lift went down to the hangar space. Maximum aircraft capacity was 22, with 11 on the flight deck.

Upgraded for action

The original secondary armament was replaced by high angle dual-purpose guns used mainly for air defence. In addition, six racks of 30-tube AA rockets were installed on sponsons aft of the flight deck. It was August 1944 before *Ise* joined Carrier Division 4 with a full complement of aircraft. By then its two Type 22 surface search radars had been upgraded and two Type 13 air search radars added. Now very much in action, the ship took part in the Leyte Gulf campaign and the Battle of Cape Engaño, shooting down five US aircraft but taking three hits. By now a target for US planes and submarines, *Ise* returned to Kure where it was used as a stationary gun platform. Towed to the base of Ondo Seto, it was under constant air attack until it was finally sunk on 28 July 1945.

Hangar

The hangar below the flight deck was fully fitted with firefighting sprays. *Ise* could carry 11 Yokosuka D4Y dive bombers plus Aichi reconnaissance planes, but never carried a full air group.

Ise underway off Sata Point, 24 August 1943, after conversion to a hybrid-aircraft carrier.

Ise

Dimensions: Length 219.68m (720ft 6in), Beam 31.75m (104ft 2in), Draught 9.45m (31ft), Displacement 40,444 tonnes (42,001 tons) full load
Propulsion: 8 oil-fired Kampon boilers, 4 sets of Kampon turbines, 4 screws, 60,000kW (80,000shp)
Armament: 4 356mm (14in) guns in twin turrets, 8 127mm (5in) guns in twin mounts, 31 250mm (1in) AA guns in twin mounts, 11 in single mounts, 6 127mm (4.5in) AA rocket launchers
Armour: Deck: 152mm (6in), 200mm (7.8in) of concrete on flight deck
Range: 14,580km (7870nm) at 16 knots (30km/h, 18mph)
Speed: 24.5 knots (45.4km/h, 28.2mph)
Complement: 1463

WORLD WAR I

Mississippi (1917)

UNITED STATES

For the *New Mexico* class, the US Navy had wanted a completely new design that would incorporate the lessons learned from its own and other navies' experience of war.

It would include 400mm (16in) guns. This was refused, but although based on the preceding *Pennsylvania* class, *Mississippi* (actually first to be commissioned) had a number of distinctive features.

Distinctive features

Most noticeable was the raked clipper-type bow that gave the ship the racy look of a hunter. The angle of the stem changed to vertical at the waterline. Oil-fired from the start, the boiler take-ups were trunked into a single funnel. The casemate arrangement for secondary guns was at last being abandoned, and during construction most of the 127mm (5in) guns were mounted at the same deck level as A turret. The remaining casemate guns were removed in the 1920s.

Armour

The armour protection was on the 'all or nothing' principle established with USS *Nevada*, with a central armoured belt 343mm (13.5in) thick at the maximum. Underwater protection consisted of a 6mm (0.25in) retaining bulkhead, an inner 76mm (3in) torpedo bulkhead and an outer 19mm (0.75in) torpedo bulkhead. Altogether, armour amounted to around 11,176 tonnes (11,000 tons) one third of the ship's displacement.

Flying-off platforms were installed on Nos. 2 and 3 superfiring turrets in 1919, and a derrick could be rigged on the stern to lift floatplanes. These rather makeshift installations were replaced in 1923 by a 24-m (78.7ft) compressed air catapult and crane on the quarterdeck, flying Vought UO-1 seaplanes. In turn this gave way to a prototype black powder catapult to test the heavier Martin MO-1 aircraft, but the ship normally carried two UO-1s and one FU-1 (the single-seat fighter version) in the 1920s. A version of this catapult remained on turret 3 until 1943.

Gun calibre

The 356mm (14in) 50 calibre gun was a developed version of the same 45 calibre guns fitted on *Nevada*. Each gun could elevate independently. In 1931 *Mississippi* was rearmed with the Mark 11 version. Twice there was a fatal flash fire in the forward superfiring turret, in 1924 and 1943, killing most of the turret crew. Each time the cause appeared to be flareback from the middle gun, igniting cordite bags stacked ready for use. Compressed air was blown through the barrels between shots to exhaust any burning debris, but burning material may have remained in the breech.

1944 profile
USS *Mississippi* around 1944.

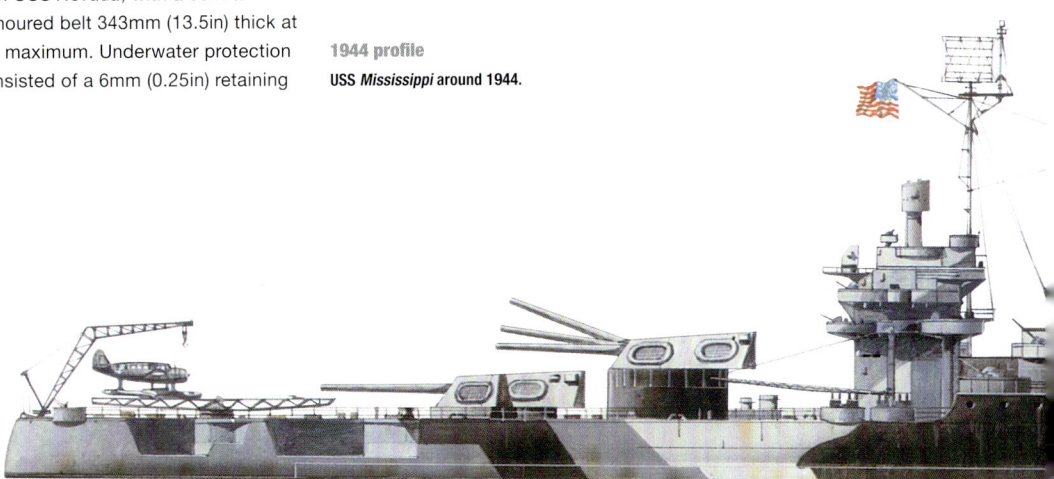

WORLD WAR I

USS *Mississippi* (BB-41) operating at sea during the late 1930s.

USS MIssissippi (after 1933)
Dimensions: Length 190.2m (624ft), Beam 29.7m (97ft 5in), Draught 9.1m (30ft), Displacement 33,960 tonnes (33,420 tons); 36,737 tonnes (36,157 tons) full load
Propulsion: 6 Bureau Express boilers, 2 sets of Westinghouse geared turbines, 4 screws, 30,000kW (40,000shp)
Armament: 12 356mm (14in) guns in triple superfiring turrets, 12 127mm (5in) guns, 8 127mm (5in) AA guns, 4 76mm (3in) guns
Armour: Belt 343–203mm (13.5–8in), Turret faces 457mm (18in), Conning tower 406mm (16in), Deck 139mm (5.5in)
Range: 11,900km (6400nm) at 10 knots (18.5km/h/11.5mph)
Speed: 22 knots (41km/h, 25mph)
Complement: 1076

WORLD WAR I

WORLD WAR I

USS *Mississippi* at New York, probably for the Victory Review of 25 December 1918.

Other modifications in the 1931–33 refit included raising of the maximum elevation of the main guns to 30°. The superstructure was built up, with gun directors for the main and 127mm (5in) guns, the funnel was heightened, the foremast taken out and not replaced, and the mainmast replaced by a short pole. Armour protection was increased, with deck armour now reaching 140mm (5.5in), capable of withstanding a 725kg (1600lb) bomb dropped from up to 1830m (6000ft), and the splinter protection above the machinery spaces was raised to 70mm (2.75in). The addition of anti-torpedo blisters was to secure underwater protection against an explosive charge of 136kg (300lb).

Mississippi's original machinery, nine Babcock & Wilcox boilers and two sets of Curtis turbines, was replaced by six Bureau Express-type boilers and Westinghouse geared turbines, pushing up its best speed to 41km/h (25mph, 22 knots). The oil bunkers had a capacity of 3331 tonnes (3279 tons). On-board aircraft consisted of three Vought 03U floatplanes, followed by Curtis SOC Seagulls in 1937 and finally Vought OS2U Kingfishers in 1941.

Battle stars

In the final years of World War I, *Mississippi* patrolled in US coastal waters. In 1920 it joined the Pacific Fleet, then with the outbreak of World War II was deployed on neutrality patrols in the Atlantic until Pearl Harbor, when it returned to the Pacific. Its speed was not high enough for fast battleship action, but it provided support for many landings and was heavily engaged in coastal bombardment. In the night Battle of Surigao Strait, 24 October 1944, its lack of up-to-date fire control radar restricted its action, but its sole broadside salvo was the last ever fired in a contest between battleships. *Mississippi* returned to the USA with eight battle stars.

In 1946, *Mississippi* was reclassified as a gunnery training ship with pennant number AG-128. With the exception of No.4 aft turret, the big guns were removed and replaced by three 127mm (5in) twin turrets, two single mount 127mm (5in) guns, and two twin 76mm (3in) guns. This armament varied in the ship's 10 years in the role. Its last operation was a test ship to launch the Terrier surface to air guided missile in 1956. It was sold for scrapping in the same year.

Leader

Class leader *New Mexico* had an experimental turbo-electric turbine drive installed. Although this had been rumoured to have also been fitted on *Mississippi*, this was not the case. Instead of using direct or geared power from turbine to propeller, it used electric power. The turbines drove one or two electric generators, which in turn powered electric motors that turned the propeller shafts. The system provided the most efficient use of turbine and propeller, enabling the turbine to keep a constant rotation rate. The electric motors were controllable by switches to increase or reduce speed, and could be reversed, obviating the need for forward and reversed turbines.

However, the system was heavier than geared turbines, more expensive to install and difficult to repair. By 1937, the Navy had reverted to geared turbines for its final battleship classes.

WORLD WAR II

In the two decades between the world wars, there were occasional naval actions but no battles. International meetings strove, with limited success, to avoid the kind of arms race that had begun in 1906. Battleship design was constrained by the Washington Treaty of 1922; great ingenuity and often subterfuge were employed in order to pack as much offensive capacity as possible into a vessel whose tonnage was limited to 35,000. Many ships were modernized and the pre-1920 ships that fought in 1939–45 were often altered beyond recognition. New elements had to be incorporated. Capital ships now carried at least two aircraft. Oil became the fuel of choice, and the possibility of refuelling at sea extended the range and endurance of ships. Rapid developments in radio communication and electronics meant frequent overhauls and replacement. Fire control was becoming steadily more sophisticated, and, along with radar, subject to great secrecy. Expertise in electronics was essential: a full and up-to-date suite of detection, defensive and offensive apparatus became as vital to a battleship as the old staples of armament, armour and machinery.

The fate of battleships
The concept of the battlecruiser had taken a pounding at Jutland, both literally and figuratively. Three of the Royal Navy's were sunk when their magazines exploded. Between Britain's HMS *Hood* in 1920 and the USS *Alaska* in 1944 only two were built, and both, USS *Lexington* and *Saratoga*, were converted into carriers, whereas the *Alaska* class was terminated with only two ships completed. The distinction between 'battlecruiser' and 'fast battleship' became very fine, since later battlecruiser designs provided more substantial armour than the earlier ones. The US 'superships' of the *Iowa* class, displacing over 50,000 tonnes (49,210 tons), combined high speed, high protection and tremendous firepower.

Awareness of the air threat to capital ships grew steadily during the 1930s, although the full intensity of sustained aerial attack was not fully appreciated until HMS *Prince of Wales* and *Repulse* were sunk on 10 December 1941. Anti-aircraft guns bristled from every possible position in the later years of the war. No new battleships were built by the United States after *Missouri* (commissioned June 1944). Britain's last, HMS *Vanguard*, was laid down in 1941 and not completed until 1946, the slow progress indicative of the priority given to submarines, destroyers and escort vessels. Even before the long-range missile and the nuclear submarine, the battleship, once the symbol of a nation's power, had been superseded by fighting units that had seemed peripheral in the days when *Dreadnought*, *South Carolina* and their ilk ruled the waves.

Ready for take-off: a Heinkel HE114 on *Gneisenau* during a fleet review in 1938.

WORLD WAR II

Hood (1920)

UNITED KINGDOM

Hood was described by one historical expert in 1968 as 'a ship far in advance of her time'. He saw it as a precursor of the fast, heavily armed battleships of World War II. However, later ships owed little to the specific features of *Hood*. It was the Royal Navy's largest warship, intended as the first of five, bigger and more heavily gunned than the German *Mackensen* class (begun but never completed). It cost just over £6 million.

Hood had a more substantial superstructure than previous ships, to accommodate a greater range of control apparatus. Fire control directors were mounted above the conning tower and on the foremast. A transmission station was in the conning tower, as was an emergency steering position.

First small-tube boilers

It was the first British capital ship to have small-tube boilers: 24 Yarrow three-drum type with a working pressure of 16.5kg/sq cm (235psi), set in four boiler rooms and providing steam to four sets of Brown-Curtis single reduction geared turbines. The gearing adjusted the revolutions of the turbines down to the rate required to turn the propeller shafts. The turbine sets were in three compartments, the forward one driving both outer propellers, the middle and aft ones driving the port and starboard propellers respectively, and each could be operated independently of the others.

Weighing the same as *Royal Sovereign*'s machinery, it gave an additional 17,897kW (24,000shp). Maximum bunker capacity was 4064.3 tonnes (4000 tons), but 1219.3 tonnes (1200 tons) was the normal load.

Focus on armour

Given that *Hood* was destroyed by shellfire from *Bismarck* and *Prinz Eugen* on 24 May 1941, much comment has focused on the extent of the ship's armour. This had already been

HMS Hood

Dimensions: Length 262m (860ft 7in), Beam 31.8m (104ft 2in), Draught 9.8m (32ft), Displacement 41,859 tonnes (41,200 tons); 45,923 tonnes (45,200 tons) full load
Propulsion: 24 Yarrow small-tube boilers, 4 Brown-Curtis geared turbines, 4 screws, 107,381kW (144,000shp)
Armament: 8 381mm (15in) guns, 12 140mm (5.5in) guns, 4 102mm (4in) AA guns, 6 533mm (21in) torpedo tubes
Armour: Belt 305–152mm (12–6in), Bulkheads 127–102mm (5–4in), Forecastle deck 50–44mm (2–1.75in), Main deck 76–19mm (3–0.75in), Upper deck 76–19mm (2–0.75in), Lower deck 76–25mm (3–1in), Barbettes 305–127mm (12–5in), Turrets 381–279mm (15–11in), Conning tower 279–229mm (11–9in)
Range: 9260km (5000nm) at 18 knots (33.3km/h, 20.8mph)
Speed: 31 knots (57.4km/h, 35.9mph)
Complement: 1433

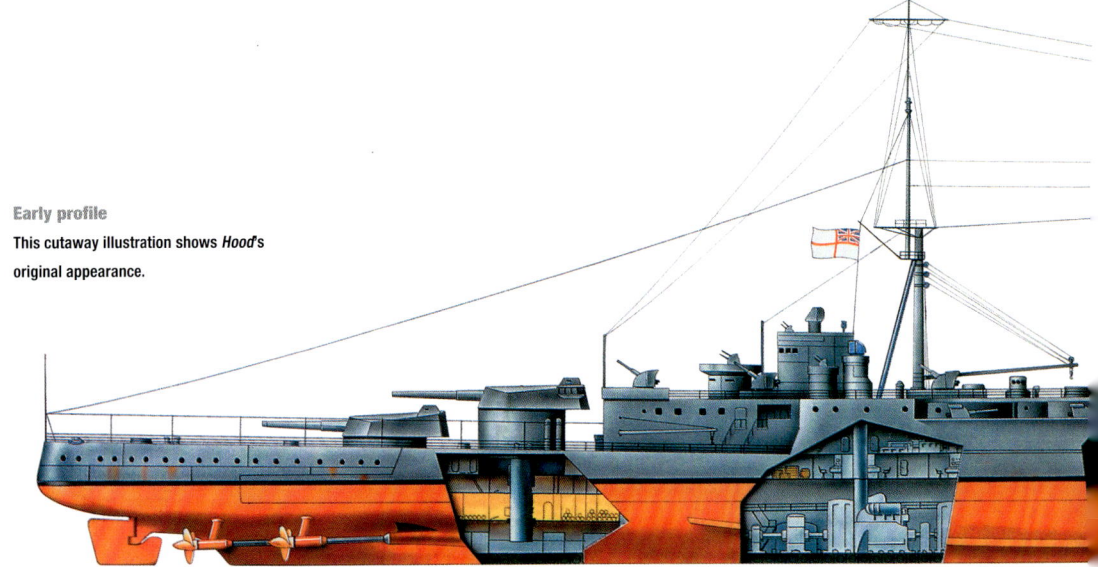

Early profile
This cutaway illustration shows *Hood*'s original appearance.

WORLD WAR II

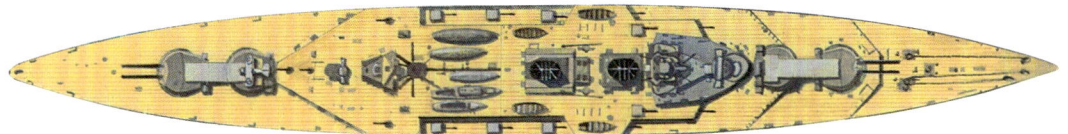

Deck plan
The deck plan shows the short-lived take-off platforms mounted on the superfiring turrets.

increased during construction (post-Jutland) and there were three armoured decks, the lowest being thickest above the machine rooms.

The side armour was graded in four bands from 305mm (12in) at the waterline to 127mm (5in) at the forecastle deck. Anti-torpedo bulges were fitted up to the top of the 305mm (12in) belt. Protection accounted for 33 per cent of the displacement, necessitated by the ship's exceptional length (*Bismarck*, shorter but wider, had 40 per cent). The rapid sinking of the ship was due to a huge explosion in the aft magazine, caused by an AP shell from one of the German ships.

From fitting out, *Hood* was equipped with the most comprehensive firing control system of any British warship. This was kept up to date in minor refits. By 1941, it carried Type 279 air warning and Type 284 gunnery radar.

Flying-off platforms were fitted to B and X turrets, for Fairey Flycatchers. A catapult was fitted on the quarterdeck in 1931 and removed in 1932.

Front view
Few modifications were made in the *Hood*'s lifetime. A major refit had been planned, but was shelved with the advent of war in 1939.

WORLD WAR II

Nagato (1920)

JAPAN

With *Nagato* the Imperial Japanese Navy (IJN) introduced the 406mm (16in) gun. Yuzuru Hiraga, its chief designer, had his sights firmly set on making the IJM a serious challenger to the US Navy (USN).

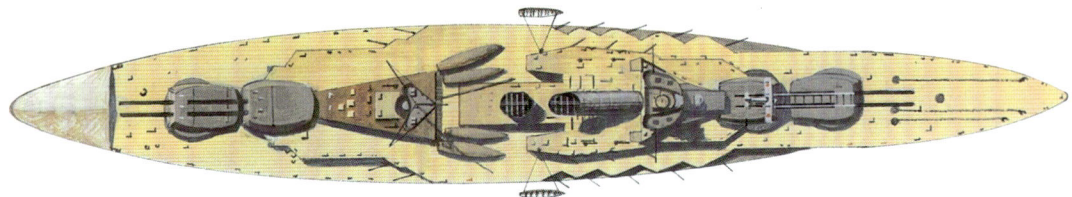

Heavily revised after Jutland, his design followed the 'all or nothing' concept, with three armoured decks above the machinery and magazines, the main one being 100mm (4in) thick and very substantial side protection including torpedo bulges outside and below the armour belt (enlarged from 1936 and partly used for additional oil storage). Twenty-one Kampon water-tube boilers (15 oil-fired, 6 coal) powered four Gihon turbines. Fuel capacity was 3455 tonnes (3400 tons) of oil and 1626 tonnes (1600 tons) of coal. In sea trials the ship reached 49.4km/h (30.7mph, 26.7 knots), considerably more than any USN capital ship.

On-board aviation began in 1925 with a flying-off platform on No. 2 superfiring turret, for Yokosuka Ro-go Ko-gata and Heinkel HD25 floatplanes. A derrick was fitted between mainmast and aft funnel in 1926 to lift planes from the sea. Further successive stages of reconstruction created its final form, with the foremast transformed into a

Nagato (pre-1935)

Dimensions: Length 215.8m (708ft), Beam 29m (95ft 3in), Draught 9.1m (29ft 9in), Displacement 32,720 tonnes (33,245 tons); 39,116 tonnes (38,500 tons) full load
Propulsion: 21 boilers (15 oil, 6 coal), 4 Gihon turbines, 4 screws, 59,656kW (80,000shp)
Armament: 8 406mm (18in) guns, 20 140mm (5.5in) guns, 4 76mm (3in) AA guns, 8 533mm (21in) torpedo tubes
Armour: Belt 305–102mm (12–4in), Deck upper 44–25mm (1.7–1in); lower 75–50mm (2.9–2.7in), Barbettes 300mm (11.8in), Turrets 356mm (14in), Conning tower forward 371mm (14.6in), aft 97mm (3.8in)
Range: 10,550km (5700nm) at 16 knots (29.6km/h, 16.25mph)
Speed: 26.75 knots (49.6km/h, 31mph)
Complement: 1333

Deck plan

Japanese naval designers paid particular attention to hull shapes, to make the ship's passage through waves as efficient as possible.

multi-platform tower, the fore funnel (curved backwards in 1925) eliminated, 10 new Kampon boilers and new turbines installed, and with new armoured main turrets enabling gun elevation to rise to 43° and extending their range to 37,900m (41,000yds). A catapult and crane were installed aft of the mainmast, now rising from an armoured control tower.

Long-range firing

From 1931, the IJN used the Type 91AP shell for long-range gunnery – 'boat-tailed', with a longer-delay fuse, anti-premature detonation device, streamlined nose cone incorporating a blunt cap-head that detached if the shell hit water. Its tri-nitro Anisol explosive charge was only 1.46–1.65 per cent of total weight. With maximum elevation, a range of 35,000m (38,800yds) could be achieved with 356mm (14in) guns, and over 38,000m (41,500yds) with 41cm (16.4in) guns. The largest version was 1.95m (77in) long, and weighed 1460kg (3219lb). This represented a change in attack procedure from short-range concentrated fire to long-range targeting of a potentially superior force.

Unique device

Fire control used a unique Japanese device, the *sokutetiban*, which computed target course and speed with an inclinometer, taking the angle between the target's course and the line of bearing towards it, and by timed changes in the bearing. By 1944, the ship mounted Type 34 fire control, Type 94 and 95 fire directors (for 127mm and 25mm (5in, 1in) for AA guns), Type 21 air search, Type 22 surface search and Type 13 early warning radars.

Philippine Sea

In 1944, *Nagato* was engaged in the battles of the Philippine Sea and Leyte Gulf, incurring some damage in the latter. Handed over to the USA in 1945, it was sunk in nuclear weapons tests in 1946.

Profile view

The illustration shows *Nagato*'s appearance in 1925–26.

Foremast

Nagato's foremast was a novel heptapod (seven-legged) structure intended to provide maximum stability for the instruments it carried. An elevator to the foretop was built into the central leg.

WORLD WAR II

Rodney (1927)

UNITED KINGDOM

With all its big guns mounted forward, *Rodney*'s design was an unusual departure for the Royal Navy. It represented the compromises necessary to make a big-gun battleship that displaced no more than 36,000 tonnes (35,000 tons) as specified by the 1922 Washington Treaty, September 1903.

Its tower bridge became the model for later British battleships and cruisers. On *Rodney* and its sister ship *Nelson*, the superstructure was shaped to enable the guns to have the widest possible arc of fire, but in practice they could fire across little more than 180° forward without causing blast damage. A short foremast was attached to the tower and a tripod mainmast had a spotting and control centre fitted below and forward of a topmast.

First three-drum boilers

Rodney and *Nelson* were among the first large warships fitted with three-drum Admiralty boilers: a compact triangular layout in which a steam drum is set on top of two water-heating drums. Water tubes form the sides of the triangle, with the oil-fired furnace in the centre. Pressure was 21kg/sq cm (300psi). The use of superheaters increased the temperature to 316°C (600°F). Eight boilers fed two sets of Brown-Curtis geared turbines driving two propellers. The turbines were set ahead of the boiler rooms, enabling the funnel to be distanced from the superstructure.

Long forward

Although its armour was substantial, including a deck 158mm (6.25in) thick over the magazine and a maximum of 356mm (14in) on the belt, efforts to keep the weight down resulted in weaknesses that required strengthening of the long forward part of the hull, which was susceptible to leaks (panting) as it flexed in the waves. There were a series of problems with the machinery.

Modifications

Few modifications were made in the ship's career, the advent of war preventing a planned major refit. A catapult was fitted to the aft turret in 1934 and a collapsible aircraft crane was added in 1937. A Fairey Swordfish torpedo bomber was replaced by a Supermarine Walrus amphibian plane. *Rodney* was the first RN battleship to acquire a radar set, a Type 79Y early warning prototype. By 1944, it had a full set of early warning, air and surface search and gunnery control radars.

Rodney's most dramatic war action came with participation in the sinking of *Bismarck* in May 1941. In reserve

WORLD WAR II

HMS Rodney

Dimensions: Length 216.8m (710ft), Beam 32.4m (106ft), Draught 9.1m (30ft), Displacement 34,493 tonnes (33,950 tons); 38,608 tonnes (38,000 tons) full load

Propulsion: 8 3-drum boilers with superheaters, Brown-Curtis geared turbines, developing 33,556kW (45,000shp), 2 screws

Armament: 9 406mm (16in) guns, 12 152mm (6in) guns, 6 119mm (4.7in) AA guns, 8 2-pounder pom-poms, 2 622mm (24.5in) torpedo tubes, underwater

Armour: Belt 356mm (14in), Bulkheads 356–76mm (14–3in), Barbettes 381mm (15in), Turret faces 406mm (16in), Deck 158mm (6.25in) max; over machinery 76mm (3in)

Range: 26,500km (14,300nm) at 12 knots (22.2km/h, 13.8mph)

Speed: 23 knots (42.6km/h, 26.5mph)

Complement: 1314

The barrels of the main nine 16-inch (406-mm) guns mounted in triple turrets on HMS *Rodney* are elevated whilst cruising the Atlantic with the Home Fleet on 10 November 1939 somewhere in the Atlantic.

Naval camouflage

The illustration shows *Rodney* in camouflage paint c. 1944. *Rodney*'s design was closely studied by other navies, with a particular influence on French practice, such as *Richelieu*.

from 1945, it was broken up in 1948–49.

The Mk1 406mm (16in) guns were of wire-wound, built-up type, and 45 calibre (i.e. the barrel length was 45 times the bore). Twenty-nine were made in order to keep *Rodney* and *Nelson* supplied with replacements. Set in triple turrets, each could be aimed individually. Weighing 109.7 tonnes (108 tons), they could elevate to 40°. Their muzzle velocity was 788m/s (2586ft/s), subsequently reduced because of excessive barrel wear. At 32° elevation the range was 32,000m (35,000yds), with AP shells weighing 929kg (2048lb). A MkII version was designed but never installed on a ship.

85

WORLD WAR II

Deutschland/Lützow (1933)

GERMANY

Strictly speaking, *Deutschland* and its sisters *Admiral Graf Spee* and *Admiral Scheer* were *Panzerschiffe* or armoured cruisers. However, the British press, with derisive intent, called them 'pocket battleships' and the name stuck.

Modern design

Laid down in 1929 at the Deutsche Werke in Kiel, *Deutschland* was commissioned in 1933, just after Hitler came to power. Although supposedly within the 10,000 imperial ton limit dictated to Germany by the Washington Treaty, it in fact displaced almost 12,000 tonnes (11,800 tons). It was the first major warship to be driven by motor engines and its modernity was evident in other ways, with a partially welded hull, and equipped from the start with provision for aircraft, gun control and range-finding equipment installed. The navigation bridge was placed quite low, behind the forward turret, with a rangefinder mounted above (the sister ships had tower-type superstructures). A pole foremast carried a searchlight platform and further sensors with a rangefinder on top. A catapult was installed in the winter of 1935–36, and two aircraft were carried: the Heinkel 60c, then the Arado 196.

Two triple turrets, fore and aft, held the six 280mm (11in) guns. Built

Deutschland/Lützow

Dimensions: Length 186m (610ft 3in), Beam 20.6m (67ft 6in) Draught 7.2m (23ft 7in), Displacement 11,938 tonnes (11,750 tons); 14,520 tonnes (14,290 tons) full load
Propulsion: 4 9-cylinder double-acting 2-stroke MAN diesel engines, Vulkan gearboxes, 2 screws, 40,268kW (54,000shp)
Armament: 6 280mm (11in) guns, 8 150mm (5.9in) guns, 3 88mm (3.5in) guns, 8 500mm (19.7in) torpedo tubes
Armour: Belt 80–60mm (3.1–2.4in), Bulkheads 45–40mm (1.8–1.6in), Deck 45–40mm (1.8–1.6in), Barbettes 150–50mm (5.9–2in), Turrets 140-85mm (5.5–3.3in), Conning tower 150–50mm (5.9–2in)
Range: 16,120km (8700nm) at 19 knots (35.2km/h, 22mph)
Speed: 28 knots (52km/h, 32mph)
Complement: 619

Paint scheme
The ship, like others, had several different camouflage paint schemes at different times during World War II.

by Krupps, they fired 339kg (739lb) AP shells to a range of 36,475m (39,890yds) at an elevation of 40°.

Built for speed

Germany had failed to make viable marine diesel engines in World War I. Such an engine was highly desirable for a ship whose weight had to be kept low, and four sets of 9-cylinder double-acting two-stroke MAN diesels (Type M9Z 42/58 7) were installed in four motor rooms. With two motors for each of two propellers, they achieved 52km/h (32mph, 28 knots), a speed that exceeded battleships of the time (this ship was intended to avoid battleships). Diesel motors also powered the four Siemens 220v electric generators, giving 2160kW (2936hp). Bunker capacity was 2750 tonnes (2710 tons).

The hull was divided into 12 watertight compartments, with a double bottom extended up to meet the armoured deck; the main bulkheads were longitudinal rather than lateral.

Commerce raider

The purpose of *Deutschland* and its sister ships was not wholly clear, but the most obvious role in a future war was as commerce raiders across the sea lanes leading to Britain. Renamed *Lützow* in 1940, *Deutschland* was used both for convoy attacks and shore bombardment in World War II, but spent much time under repair after air attacks, until sinking in shallow water at Swinemünde on 4 May 1945.

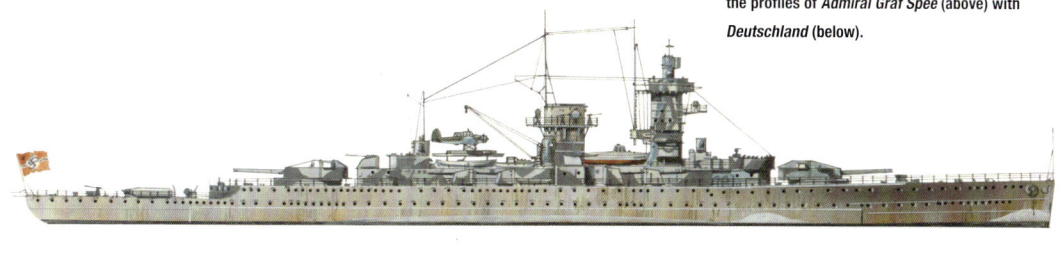

Admiral Graf Spee
The pocket battleships were not identical: compare the profiles of *Admiral Graf Spee* (above) with *Deutschland* (below).

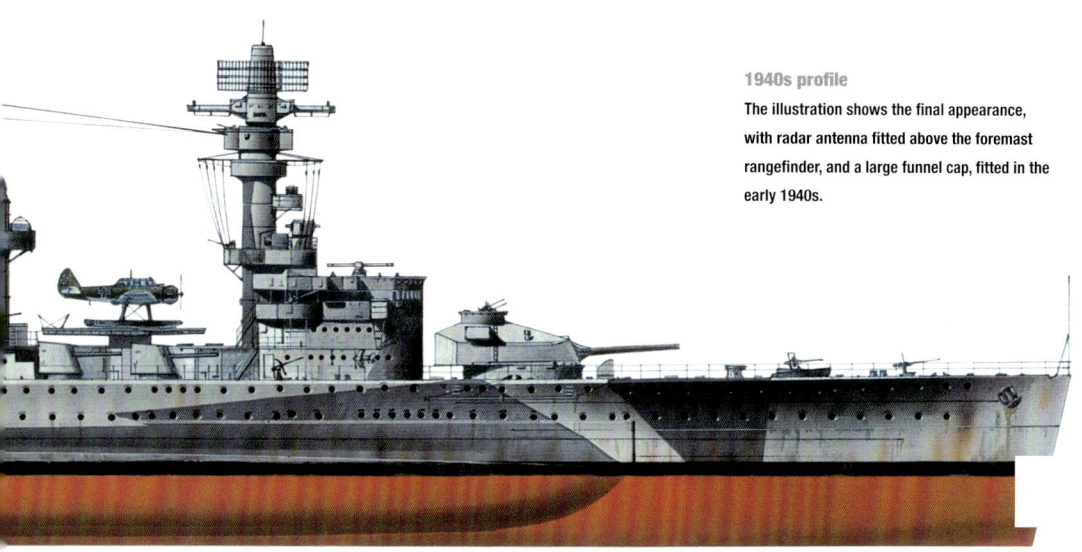

1940s profile
The illustration shows the final appearance, with radar antenna fitted above the foremast rangefinder, and a large funnel cap, fitted in the early 1940s.

WORLD WAR II

Dunkerque (1937)

FRANCE

Laid down at Brest on 24 December 1932, *Dunkerque* was not liable to the same Treaty regulations as German ships. Its larger guns and greater speed were a response to the *Deutschland* class. The Italians followed suit and designed new ships with even larger guns. It was the leapfrog game of 1906–18 being played again, although with fewer ships involved.

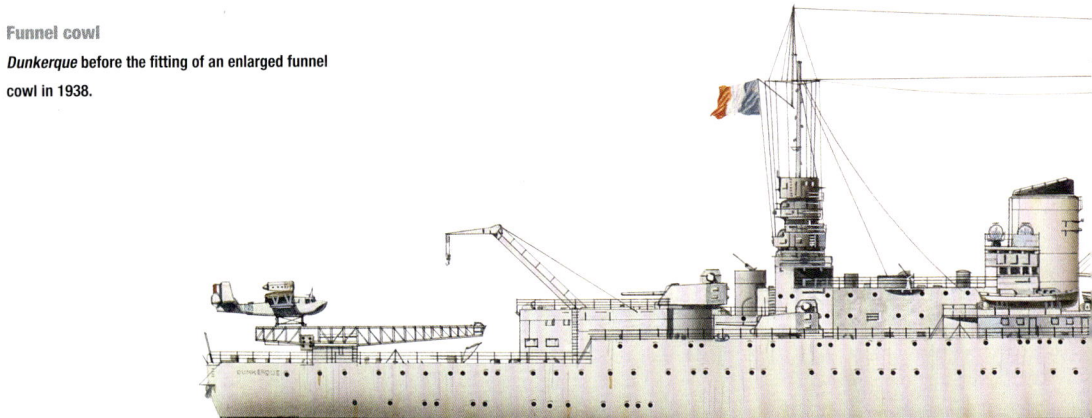

Funnel cowl
Dunkerque before the fitting of an enlarged funnel cowl in 1938.

Dunkerque's superstructure followed the form established with HMS *Rodney*, a tower-cum-mast rising behind the armoured conning tower, which was topped by the navigation bridge. In the tower were an admiral's bridge, directors for the main and secondary guns, searchlight platforms and radio aerials, and a 12m (39ft 4in) rangefinder. An interior lift was fitted. The control stations were gas-proof and flash-proof. The pole mainmast was built up around directors for the aft guns.

Armour and guns
Again, like *Rodney*, the main guns were all forward, but set in two quadruple turrets – a novel arrangement. To avoid the whole turret being knocked out by a hit, each was divided in two by a 45–25mm (1.8–1in) armoured wall; and the two turrets were set 27m (88.6ft) apart.

The main turrets, built by St-Chamond, each weighed 1497 tonnes (1473 tons). The backs, extending beyond the barbettes, were heavily plated, partly to balance the guns: 345mm (13.6in) on the forward one, 335mm (13.2in) on the superfiring one. Elevating to a maximum 35°, the guns had a muzzle velocity of 870m/s (2854ft/s) and a range of 41,500m (45,385yds).

The shells weighed 570kg (1256.6lb), almost twice as much as *Deutschland*'s 280mm (11in) shells, and a third more than the 305mm shells of Italian battleships.

A 22m (72ft) compressed-air catapult and a crane were fitted on the after deck, aft of a two-level hangar, with lift, and three Loire 130 seaplanes were carried.

The side armour covered about 60 per cent of the hull, with bow and stern unprotected. A 60mm (2.4in) layer of teak wood backed the outer plating. Internally, longitudinal and lateral bulkheads framed the machinery space, and two armoured decks were installed to counter plunging fire, varying from 125mm (4.9in) to 40mm (1.6in).

Unusual boiler layout
Six Indret superheated, water-tube boilers were installed in three rooms, powering four sets of Parsons geared turbines in two rooms: in an unusual arrangement, the forward boiler room was separated from the other two by the first turbine room, with the second turbine room placed aft of boiler room 3. This provided two separate drive systems that could operate independently in the event of damage.

WORLD WAR II

Dunkerque

Dimensions: Length 214.5m (704ft), Beam 31.1 (102ft), Draught 8.7m (29ft), Displacement 26,925 tonnes (26,500 tons); 36,070 tonnes (35,500 tons) full load
Propulsion: 6 Indret boilers, 4 sets of Parsons geared turbines, 4 screws, 80,200kW (107,500shp)
Armament: 8 330mm (13in) guns, 16 130mm (5.1in) guns, 8 37mm (1.5in) guns, 32 13.2mm (0.5in) AA guns
Armour: Belt 225mm (8.9in), Main deck 115mm (4.5in), Barbettes 310mm (12.2in), Turrets 345–310mm (13.6–12.2in), Conning tower 270mm (10.6in)
Range: 14,540km (7350nm)
Speed: 31 knots (57.5km/h, 35.7mph)
Complement: 1432

The 330mm (13in) guns were mounted in individual cradles with a maximum elevation of 350°.

The exhaust from boiler room 1 was trunked back in a double-bend pipe. A diesel engine was fitted on the bottom deck between the two forward turrets to drive the electricity generators.

In 1938, a funnel cap was fitted to help keep smoke away from the control stations. Fire control and optics were largely French-made, from OPL (Optique et Précision de Levallois-Perret); the three stations on the tower weighed a total of 90 tonnes (88.5 tons).

In World War II, *Dunkerque* was flagship of the Atlantic fleet at Brest before transfer to the Mediterranean in 1940. It was damaged in the British attack on Mers-el-Kebir on 3 and 6 July that year. Moored facing the land, it could not return fire. Transferred to Toulon, it was scuttled by its crew in November 1944.

WORLD WAR II

Gneisenau (1938)

GERMANY

Was this ship, and its sister *Scharnhorst*, a battleship or a battlecruiser? The question arises primarily because their 280mm (11in) guns were outmatched by the battleships of other nations.

Their roles, too, were more those of a fast heavy cruiser rather than joining the 'line of battle' against other capital ships. Yet that was an outmoded tactic, at least in the North Sea and the Atlantic, by the mid-1930s. To the Kriegsmarine, they were *Schlachtschiffe*, battleships.

Gneisenau was laid down at the Deutsche Werke in Kiel in March 1935 and commissioned in May 1938. The profile was distinctive, with a control tower set well back behind an armoured deckhouse, a foremast attached abaft the single funnel and a short mainmast set on the aft control platform. The main guns were in triple turrets, two forward and one aft.

The drive machinery comprised 12 Wagner high pressure (58kg/sq cm,

Gneisenau
Dimensions: Length 229.8m (753ft 11in), Beam 30m (90ft 5in), Draught 8.23m (27ft), Displacement 35,397 tonnes (34,840 tons); 39,522 tonnes (38,900 tons) full load
Propulsion: 12 Wagner HP boilers, 3 Germania geared turbines, 3 screws, 123,040kW (165,000shp)
Armament 9 280mm (11in) guns, 12 150mm (5.9in) guns, 14 105mm (4.1in) guns, 16 37mm and 8 20mm AA guns
Armour: Belt 350–200mm (13.8–7.9in), Bulkheads 200–150mm (7.9–5.9in), Deck 50–20mm (2-0.8in), Barbettes 350–200mm (13.8–7.9in), Turrets 350–200mm (13.8–7.9in), Conning tower 350-100mm (13.8–3.9in)
Range: 16,297km (8,800nm) at 19 knots (35km/h, 21.8mph)
Speed: 32 knots (59.2km/h, 32.5mph)
Complement: 1840

Combined, the *Gneisenau*'s 280mm (11in) guns could fire 63 shells in two minutes.

WORLD WAR II

825psi) water-tube boilers and three sets of Krupp Germania turbines: for this much larger and faster ship the diesel engines of *Deutschland* would not have sufficed. Five sets of diesel-powered generators were positioned around the ship. Both ships had a bunker capacity of 2800 tonnes (2755 tons), but using voids between the belt armour and the torpedo bulkhead, a further 2000 tonnes (1968 tons) could be carried.

Built in a hurry

One reason for the 280mm (11in) guns was that the ships were built in something of a hurry. No bigger gun had yet been developed and tested, and the barbettes were theoretically convertible to hold heavier (twin 380mm/15in) gun turrets as and when they became available. As fitted, the guns were improved versions of *Deutschland*'s guns, with a higher calibre (34 compared to 28.3) and reinforced barrels. They fired HE or AP shells weighing 300kg (661.4lb). With a muzzle velocity of 910m/s (2986ft/s) and an elevation of up to 40°, the range extended to 36,475m (39,890yds), with a maximum for the superfiring 'Bruno' turret of 40,930m (44,760yds). Seven rounds could be fired in two minutes, and each ship carried 900 shells. The turrets traversed electrically, but elevation was hydraulic.

Armour was of Krupp hardened steel in different forms: KCA in the central belt, the more ductile '*Wotan weich*' (soft) on places where oblique hits could be expected, and 'Wotan hart' on the most vital spots. Central anti-torpedo protection, below the belt, had an outer skin from 66–12mm (2.6–0.47in), and a watertight void with an 8mm (.31in) between it and the fuel bunkers. Longitudinal and transverse bulkheads provided strength. This scheme was intended to withstand a hit from a 250kg (500lb) explosive warhead.

Germany was ahead of other naval powers in its use of solventless propellants in gunnery. It is notable that when *Gneisenau* was bombed at Kiel in 1942, 24 tonnes (23 tons) of RP C/32 shell propellant in a magazine below 'Anton' turret was ignited but did not explode, although the bomb blast lifted the 750 tonnes (738 tons) some 50cm (20in).

Float plane

Gneisenau carried three Arado Ar 196A aircraft. The catapult on the aft turret had been removed by 1941.

WORLD WAR II

Scharnhorst (1939)

GERMANY

As the specifications and images show, *Scharnhorst*, built at Deutsche Werke, Wilhelmshaven, differed from *Gneisenau* in certain details. Both ships were built with a straight stem, which proved ineffective in deflecting waves, and was replaced in 1939 with the raking 'Atlantic' bow form, which was more seaworthy and enhanced their appearance.

Scharnhorst had a foremast attached to the control tower and a pole mainmast forward of the aft control tower, with searchlight and spotting platforms. Other early modifications included the midships hangar and funnel caps. After 1942 two triple sets of deck-mounted 533mm (21.3in) torpedo tubes were installed. Both ships were constructed on longitudinal steel frames, with transverse bulkheads forming 21 watertight compartments below the waterline. Welding was extensively used in building the hull, prior to the bolting on of the side armour. A cellular double bottom made up 80 per cent of the keel length.

Armament
Main and secondary armament were the same on both, the lighter guns comprising 12 150mm (5.9in) guns, alongside the superstructure – 8 in twin turrets and 4 single-mounted. A large AA battery was installed: 14 105mm (4.1in) in twin mounts and 16 37mm (1.5in) also in twin mounts. Maximum elevation was 80° (105mm/4.1in; 85° for 37mm/1.4in) enabling close range AA fire, with a ceiling of 6800m (15,750ft).

Rangefinders and high-quality Zeiss optics were supplemented by *Seetakt* radar, version *Dete1*, with a maximum range of 220km (140mi). One set was mounted on the forward

Scharnhorst
Dimension: Length 235m (772ft) Beam 30m (98ft 5in), Draught 9.7m (31ft 9in), Displacement 32,615 tonnes (32,100 tons); 38,711 tonnes (38,100 tons) full load
Propulsion: 12 Wagner HP boilers, 3 Brown-Boveri geared turbines, 3 screws, 119,312kW (160,000shp)
Armament: 9 280mm (11in) guns, 12 150mm (5.9in) guns, 14 105mm (4.1in) guns, 16 37mm (1.5in) and 10 20mm (0.79in) AA guns, 6 533mm (21.3in) torpedo tubes
Armour: Belt 350–200mm (13.8–7.9in), Bulkheads 200-150mm (7.9–5.9in), Deck 50–20mm (2--0.8in), Barbettes 350–200mm (13.8–7.9in), Turrets 350–200mm (13.8–7.9in), Conning tower 350-100mm (13.8–3.9in)
Range: 18,500km (10,000nm) at 17 knots (31.5km/h, 19.7mph)
Speed: 31 knots (57km/h, 35.6mph)
Complement: 1669

Late profile
Scharnhorst post-1942, with the catapult on the aft turret removed.

gun director (on the bridge), the other on the aft main battery director, showing that it was chiefly used for ranging rather than for combat in darkness or fog.

On both, hangar space was built into the aft superstructure, with a catapult on top and crane alongside. A catapult was also placed on *Scharnhorst*'s aft main turret between 1939–42. Three Arado Ar 196A aircraft were carried.

Dramatic history

Both ships had dramatic histories. On several occasions, their superior speed saved them from encounters with more heavily armed battleships, but *Scharnhorst*'s end came on 26 December 1943, under the 356mm (14in) guns of the RN battleship *Duke of York*, with gunfire and torpedoes from a supporting force of cruisers and destroyers.

Salvo effect

With a muzzle velocity of 890m/s (2900ft/s), the 280mm (11in) shells could penetrate 291mm (11.5in) of side armour at a range of 18,288m (20,000yds). This was the distance from which *Scharnhorst* and *Gneisenau* opened fire to sink the carrier HMS *Glorious* on 8 June 1940.

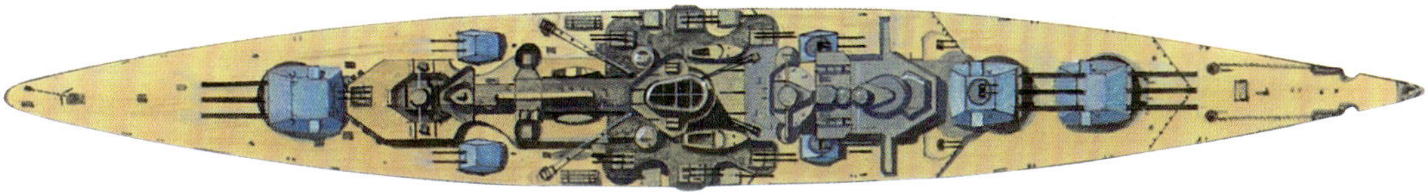

WORLD WAR II
Bismarck (1940)

GERMANY

Bismarck **and its sister *Tirpitz* were the largest and most powerful battleships yet built. Only *Yamato*, still under construction in 1940, and the USN *Iowa* class, would surpass them.**

Bismarck was laid down at Blohm & Voss, Hamburg, on 1 July 1936, launched on 14 February 1939 and completed on 24 August 1940. Its cost was 196,800,000 Reichsmarks. The chief designer, Dr Hermann Burkhardt, head of the Naval Construction Office, was also responsible for *Gneisenau* and *Scharnhorst*, as part of an ongoing programme that would enable the Kriegsmarine to match the Royal Navy by 1950.

Variations in detail
In appearance the ship was an enlarged *Scharnhorst*, although with many differences of detail. The bow was remodelled after launching but

Adolf Hitler inspects KMS *Bismarck* at Gdynia, 5 May 1941.

Obselete design
It has been suggested that *Bismarck*'s design was obsolete in some respects, especially the armour distribution. But the ship's fate was sealed by torpedo damage to the rudders.

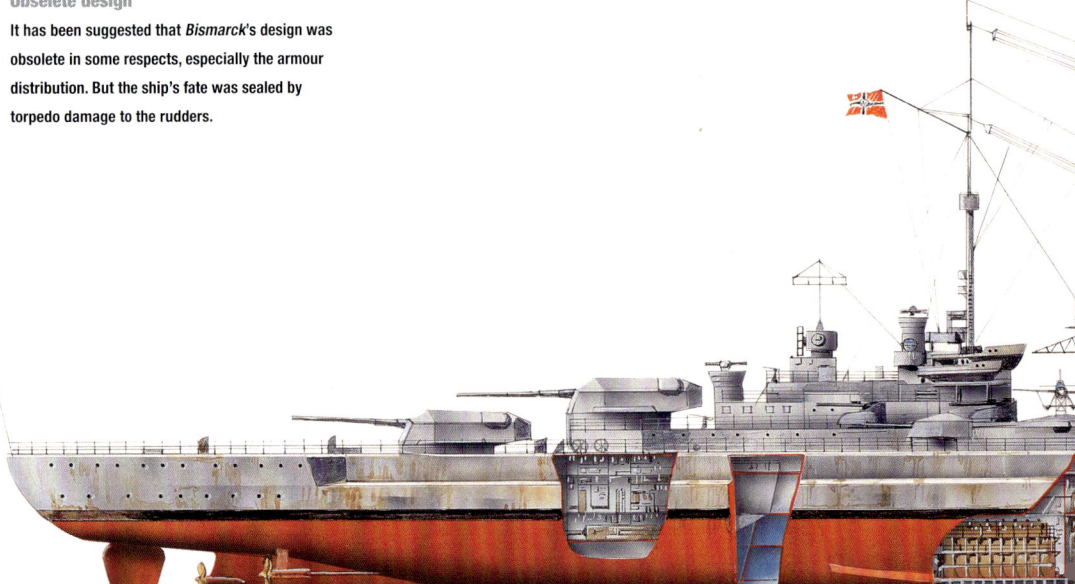

WORLD WAR II

before completion, to the 'Atlantic' form as fitted to *Scharnhorst*. The control tower was set well back from the navigation bridge and conning tower. The foremast, no more than an aerial pole, was attached on the aft end. The base of the single oblong funnel sloped forward to accommodate the flues from the forward boiler rooms. The funnel itself, with a tilted cap, was joined to the tower by a flying bridge and ringed by a platform with a 1.5m (59in) diameter searchlight at each angle, the forward lights having protective caps. A crane jib was fixed to the funnel's aft side. A pole mainmast with two topmast sections was fitted forward of the aft rangefinder tower. The aircraft catapult was a lateral one between the two superstructure houses.

The hull was constructed from standard ships' plating ST52 steel, and more than 90 per cent was of electric welding construction. For Burkhardt, the stability of a battleship as a firing platform was a key consideration and *Bismarck*'s beam was 36m (118ft) at its widest, a feature that also enabled a high degree of internal protection. From foretop to keel were 17 levels or decks. Internally the ship was divided into 22 watertight compartments – 17 within the armoured zone, which accounted for 70 per cent of the ship's waterline length.

Gun power

The main guns were eight 380mm (15in) mounted in twin superfiring turrets, arranged to keep the guns as far apart as possible to avoid shocks and interferences. These guns, of 51.66 calibre, had been under development since 1934. Their

Bismarck

Dimensions: Length 251m (793ft), Beam 36m (118ft), Draught 9.3m (31ft), Displacement 41,700 tonnes (41,000 tons); 50,300 tonnes (49,500 tons) full load
Propulsion: 12 Wagner HP boilers, 3 Brown-Boveri geared turbines, 3 screws, 111,982kW (150,170shp)
Armament: 8 380mm (15in) guns, 12 150mm (5.9in) guns, 16 105mm (4.1in) guns, 16 37mm (1.5in) guns, 12 20mm (0.79in) AA guns
Armour: Belt 320–80mm (12.6–3.1in), Bulkheads 220–45mm (8.6–1.7in), Deck 120–80mm (4.3–1.7in), Barbettes 340–220mm (13.4–8.6in), Turrets 360–180mm (14.2–7in)
Range: 16,430km (8870nm) at 19 knots (35km/h, 22mph)
Speed: 29 knots (53.7km/h, 33.4mph)
Complement: 2221

WORLD WAR II

weight was 109.2 tonnes (107.5 tons). Range was 36,200m (39,589yds) at an elevation of 35° (giving a low trajectory and shortening shell flight time). Its best rate of fire was one round every 20 seconds. The shells weighed 798kg (1,759lb) and were of three kinds: armour-piercing HE for use against battleships, and two types of HE shell, one with a head fuse and one with a base fuse.

Twelve 150mm (5.9in) guns were fitted in twin turrets, three on each side. AA defences consisted of 16 105mm (4.1in) twin-mount guns, 16 37mm (1.5in) twin-mount and 12 20mm (0.79in) single-barrel guns. Gunnery control was located in three positions: on the forward conning tower, above the foretop platform and on the rear conning tower, with communication links to two computation rooms (within the armoured zone fore and aft). *Bismarck* carried no torpedoes, but both ships had four Arado Ar196 floatplanes.

Krupp armour

Armour was Krupp-made. The company had continued to develop new defences to counter new and larger shell and torpedo types, and *Bismarck* had three kinds. Krupp KCn/A, a face-hardened version of the Krupp cement type, for the belt, turrets and control towers; Wh – *Wotan hart* ('hard') – homogeneous armour steel, for the armoured decks, and Ww – *Wotan weich* ('soft') – with a lower tensile strength, for the longitudinal torpedo bulkheads. Altogether the armour weight was 17,533 tonnes (17,256 tons), 32.2 per cent of the ship's total displacement.

The machinery comprised 12 Wagner high-pressure boilers in six watertight compartments, providing high-pressure (58kg/sq cm/825psi) steam at 450°C (842°F) to three sets of Curtis-type single reduction geared turbines, each in separate rooms and driving the three screws. Each set had high pressure (HP), intermediate pressure (IP) and low pressure (LP) turbines. Unlike *Tirpitz*, the ship had no cruising turbine fitted. Eight 500kW diesel generators, five 690kW and one 460kW turbo-generators were installed in four plants. The regular bunkers stored 3282 tonnes (3230 tons) of oil but use of other void spaces could double this.

Bismarck was a hard ship to sink. After the sinking of *Hood*, almost

Bismarck at sea, seen from *Prinz Eugen*, 18 May 1941.

succeeding in dodging the pursuit, it was hit by Swordfish torpedo bombers, jamming the port rudder at 120° to port, but kept up a vigorous defence.

On 27 May at 08.47, battleships HMS *Rodney* and *King George V* opened fire with 406mm (16in) and 356mm (14in) guns at around 20,000m (21,900yds), gradually closing to 7860m (8600yds). By around 10.00 all *Bismarck*'s guns were disabled and the order was given to scuttle. Hit by more torpedoes, it capsized and sank at 10.39, with 1977 of the 2221 crew on board.

WORLD WAR II

Richelieu (1940)

FRANCE

Basic plans for *Richelieu* were drawn in the early 1930s. Laid down at Brest in 1935, its design underwent numerous modifications before the launch on 17 January 1939 and commissioning on 15 June 1940.

Early profile
Richelieu prior to the 1943 refit.

It was a substantially enlarged version of the *Dunkerque*, with a similar hull form although almost 40m (131.2ft) longer. It had a raised central deck area, the main guns all forward in two quadruple turrets, and the same design of conning and command tower, with three rangefinders on the same central axis.

The aft section of superstructure was quite different, with the box-shaped funnel combined in the aft control tower, tilted backwards at an angle of 45°. Masts were short toppings to the command towers, although the foremast was heightened in 1943. As with *Dunkerque*, two catapults were mounted at the stern, with hangar space for four Loire 130 aircraft.

Its long hull made for speed, and power came from six Indret forced-circulation (internal pump) boilers, driving four sets of Parsons geared turbines. The machinery was arranged in successive rooms, fore to aft: boiler room 1/turbines for outer screws/boiler room 2/turbines for inboard screws, with the aim of ensuring mobility even if one set were put out of action. Maximum bunker capacity was 5886 tonnes (5773 tons) of oil.

The eight 350mm (15in) guns were housed in two separate compartments inside each turret, with an armoured bulkhead between. They could elevate to 35°, with a muzzle velocity of 830m/s (2700ft/s). At 30°, the range was 34,800m (38,058yds). The shells were 884kg (1949lb). Secondary armament was nine 152mm (6in) dual-purpose guns in triple turrets, placed aft, one on the central axis, two lateral. Twelve 100mm (3.9in) and 12 37mm (1.5in) AA guns were fitted, along with 32 13.2mm machine guns.

Total armour weight was 16,045 tonnes (15,788 tons). By comparison, *King George V* had 12,500 tonnes (12,300 tons) and *Bismarck* had 17,256 tonnes (16,980 tons), although differently disposed in each case.

1943 refit
Held by the Vichy regime after the fall of France in 1940, *Richelieu* exchanged fire with RN ships at Dakar in July 1940. In 1942, it was taken over by Free French forces, and got a refit in New York in 1943. Three guns in the superfiring turret were replaced and 50 Oerlikon Mk4 20mm (0.75in) AA guns were installed in place of the aviation equipment and on the foredeck. Fourteen Bofors 40mm (1.5in) AA guns were placed around the superstructure

WORLD WAR II

Richelieu

Dimensions: Length 247.9m (813ft), Beam 33m (108ft), Draught 9.7m (32ft), Displacement 43,987 tonnes (43,293 tons); 48,311 tonnes (47,548 tons) full load

Propulsion: 6 Indret superheated boilers, four Parsons geared turbines, 4 screws, 111,855kW (150,000shp)

Armament: 8 380mm (15in) guns in quadruple mounts, 9 152mm (6in) guns, 12 100mm (3.9in) AA guns, 14 37mm (1.5in) AA guns

Armour: Belt 343–250mm (13.5–9.8in), Bulkheads 383–251mm (15.1–9.9in), Deck upper 170-130mm (6.7–5.1in), Deck lower 100–40mm (3.9i–1.6in), Barbettes 405mm (16in), Turrets 430-195mm (16.9–7.7in)

Range: 18,240km (9,850nm) at 16 knots (30km/h, 18.75mph)

Speed: 30 knots (56km/h, 35mph)

Complement: 1550

The compact design of the boilers did not reduce their efficiency. On trials the ship reached 60.4km/h (37.5mph/32.6knots) indicating a power output of 131,654kW (179,000shp).

Side belt

Richelieu's side belt went deeper than its contemporaries: 7.01m (23ft), compared to 6.09m (20ft) on Bismarck and 5.46m (17.9ft) on Iowa.

and forward turrets. SF and SA2 radar equipment was installed. Richelieu's value to the Allies was curtailed by the unavailability of shells for its main guns, but it served in the Far East and the Mediterranean. After World War II it remained active until 1959. It was scrapped in 1968–69.

WORLD WAR II

Yamato (1940)

JAPAN

Yuzuru Hiraga was influential in the design of *Yamato* and *Musashi* – what were to be the world's biggest warships until the advent of the American super-carriers later in the century.

In 1933 the Imperial Japanese Navy's arsenal at Kure produced experimental guns of 460mm (18.1in) calibre. A full battery of these would have required a battleship considerably larger than any yet built – and far beyond the maximum allowed by the London Treaty. When in December 1934 the Japanese government withdrew from the Treaty, planning had already begun, in top secret, for four super-battleships that would outclass and outfight any other warship. They would form the nucleus of an unbeatable battlefleet. With nine Type 94 460mm (18.1in) guns, they would have armour capable of withstanding 18in shells fired from 20,000 to 30,000m, a top speed of 30 knots and a cruising range of 12,800km (8000 mi) at 18 knots. Around 23 successive plans were drawn up before the final design for *Yamato* and *Musashi* was approved in March 1937.

Distinctive appearance

Striking aspects of the ship's appearance included the typical Japanese tall tower, more compact than on converted older ships, and a single large backwards-tilted funnel. There was no foremast and an aft-leaning mainmast set on a tripod. It was flush-decked almost as far aft as the stern, where a catapult and aircraft crane were mounted.

Extensive testing resulted in a bow design with a narrow stem enlarging into a bulbous underwater projection, widening to a (38.9m, 127ft 7in) beam

Deck plan
The plan view shows the strongly centre-line configuration on a spacious deck area.

broad in relation to the waterline length of 256m (839ft 11in), allowing for six 155mm (6.1in) guns to fire forward as well as the main guns. Another notable feature was the downwards slope of the foredeck between the turrets, enabling the main guns to be set slightly lower. Its width helped sustain the stresses when the huge guns were fired, and also ensured a good degree of stability. Despite its size, *Yamato*'s manoeuvring ability was excellent: an important consideration in torpedo evasion.

The wide beam enabled its Kampon boilers to be set in four longitudinal rows of three, each row serving four Kampon turbines, reducing the machine area to 53 per cent of waterline length, with a consequent saving in the extent of armour protection. Four 6m (20ft) propellers were fitted. The ship achieved its intended speed, but comparison of *Yamato*'s and USS *Iowa*'s machinery shows some striking differences. The Japanese ship had 12 boilers, while the American had eight.

Main guns
Yamato's main triple-mount turrets each weighed 2818 tonnes (2774 tons)

BATTLESHIPS COMPARED
Boiler steam pressure: *Yamato* 25kg/sq cm (355.6psi), *Iowa* 39.7kg/sq cm (565psi)
Steam temperature: *Yamato* 325°C (618°F), *Iowa* 454°C (850°F)
Power output: *Yamato* 110,325kW (147,948shp), *Iowa* 158,088kW (212,000shp)
Machinery weight: *Yamato* 6599 tonnes (6495 tons), *Iowa* 5980 tonnes (5980 tons)
Power output per tonne: *Yamato* 16.7kW (22.4shp), *Iowa* 26.4kW (35.4shp)
Maximum speed: *Yamato* 27 knots, *Iowa* 33 knots

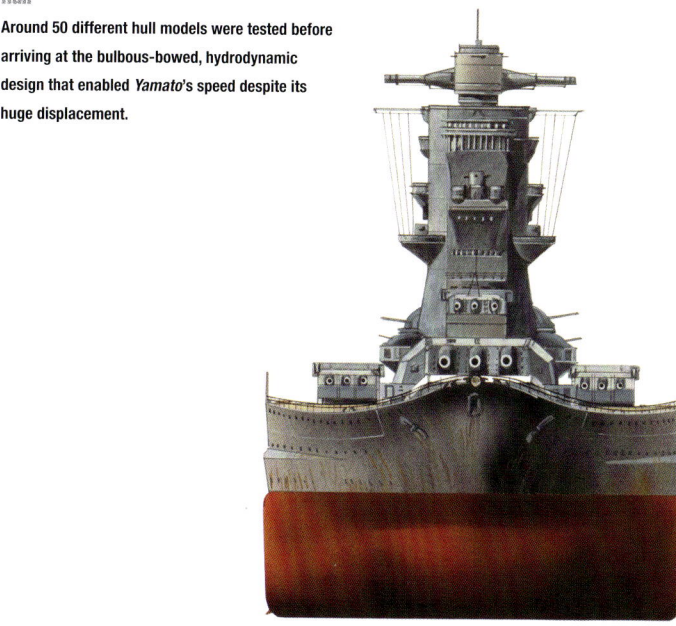

Hull
Around 50 different hull models were tested before arriving at the bulbous-bowed, hydrodynamic design that enabled *Yamato*'s speed despite its huge displacement.

with separate compartments for each gun. Their front armour was 558.8mm (22in) thick, with 406.4mm (16in) on the sides, of specially hardened steel. The gun barrels were 21.13m (69ft 4in) long, each 45 calibre. They could elevate from -5° to 45°, and had a muzzle velocity of 780m/s (2600ft/s). Maximum firing range was 42,000m (45,932yds), although they were most effective at 25,000m (27,340yds). They fired AP Type 91 shells of 1460kg (3218.7lb) and HE and AA shells of 1360kg (2998.3lb).

Under optimum conditions a rate of fire of one round every 30 seconds was achieved. Total magazine capacity was 560 shells – 60 per gun. The USN reckoned that the 406mm (16in) guns of the *Iowa* class were as effective in armour piercing as the 460mm (18.1in) Japanese gun, despite the latter's greater weight and higher muzzle velocity, owing to a better design of shell, with greater frontal density and consequent greater impact than the Japanese type.

As secondary armament, four triple turrets held 12 155mm (6.1in) guns.

The beam turrets were removed in the course of the war, to make way for more AA guns. By the end of its short career the ship was fitted with 288 25mm AA guns, compared with 24 in the original design.

Two catapults were mounted at the stern, with a hangar below. Seven aircraft could be carried: in 1942 they were four Aichi E13As and three Mitsubishi F1Ms. The ship's boats were also held in enclosed hangars, with a mechanized rail system to remove and launch them.

Strong armour

Yamato's armour was very strong where it mattered, as the first Japanese warship to follow the 'all or nothing' principle. The armoured box enclosing the machinery and magazines accounted for 55 per cent of the ship's length, and was designed to resist 460mm (18.1in) shellfire at ranges between 19,930–29,930m (21,800–32,800yds) as well as 1000kg (2200lb) bombs dropped from up to 3400m (11,150ft). The total weight was almost 23,370 tonnes (23,000 tons). Outside the heavy armour there were over 1000 separate internal compartments. There were flaws, however. Compared to *Bismarck*, *Yamato* and *Musashi*'s provision of watertight compartments was inadequate, especially in the lightly protected bow areas. The armour joints were defective.

No feasible amount of armour, however, could have sustained the ship against the intensive attacks of US torpedo bombers on 7 April 1945. Struck by at least 11 torpedoes and six bombs, it capsized and sank.

Yamato
Dimensions: Length 263m (862ft 9in), Beam 36.9m (121ft 1in), Draught 10.39m (34ft 1in), Displacement 69,098 tonnes (68,010 tons); 72,806 tonnes (71,659 tons) full load
Propulsion: 12 Kampon HP boilers, 4 Kampon turbines, 4 screws, 111,855kW (150,000shp)
Armament: 9 460mm (18.1in) guns, 12 155mm (6.1in) guns, 12 127mm (5in) guns, 24 25mm (1in) and 4 13.2mm (0.52in) AA guns
Armour: Belt 410mm (16in), Deck 230–200mm (9.1–7.9in), Barbettes 546–50mm (21.5–2in), Turrets 650–193mm (19.7–11.8in), Torpedo Bulkhead 300–75mm (11.8–2.9in)
Range: 13,330km (7200nm) at 16 knots (29.6km/h, 18.5mph)
Speed: 27 knots (49.9km/h, 31.2mph)
Complement: 2500

Yamato under air attack as a bomb explodes off its port side, 7 May 1945. The fire in the area of the 155mm (6.1in) turret can be clearly seen.

WORLD WAR II

Vittorio Veneto (1940)

ITALY

Although launched in July 1937, *Vittorio Veneto* did not enter service until completion in May 1940. It was built at Trieste by Cantieri Reuniti dell'Adriatico, simultaneously with the almost identical *Littorio* (built at Genoa).

The design group was headed by Umberto Pugliese, general inspector of the Naval Engineer Corps. Intended to counter the French *Dunkerque* class, these were fast and powerful battleships. Navigation, command and fire control were all incorporated in the tower, with no forward superstructure. The pole foremast was linked to the tower by three flying bridges, and the short mainmast was fixed to the aft control tower.

Italian-designed Belluzzo turbines were powered by eight Yarrow boilers. The four turbine sets were in independent compartments, two forward of the boilers, and two aft.

Ingenious protection

The armour included an outer layer of decapping plates, intended to break off the penetrative caps fitted to AP shells. With a thickness of 70mm (2.8in) it could decap shells of up to 380mm (15in) – the calibre of *Richelieu* and *Dunkerque*'s main batteries. *Vittorio Veneto* and *Littorio* both incorporated Pugliese's ingenious torpedo protection system, formed of longitudinal cylinders within the ship's side, packed with tubes, intended to absorb the shock of a torpedo hit. On their inner side a longitudinal watertight bulkhead was intended to protect the machinery and magazines. It avoided the torpedo bulges that reduced the maximum speed of other battleships, but was criticized for the weakness of the inner bulkhead.

The Ansaldo M134 381mm (15in) main battery was capable of firing a salvo of 885kg (1951lb) shells at the high

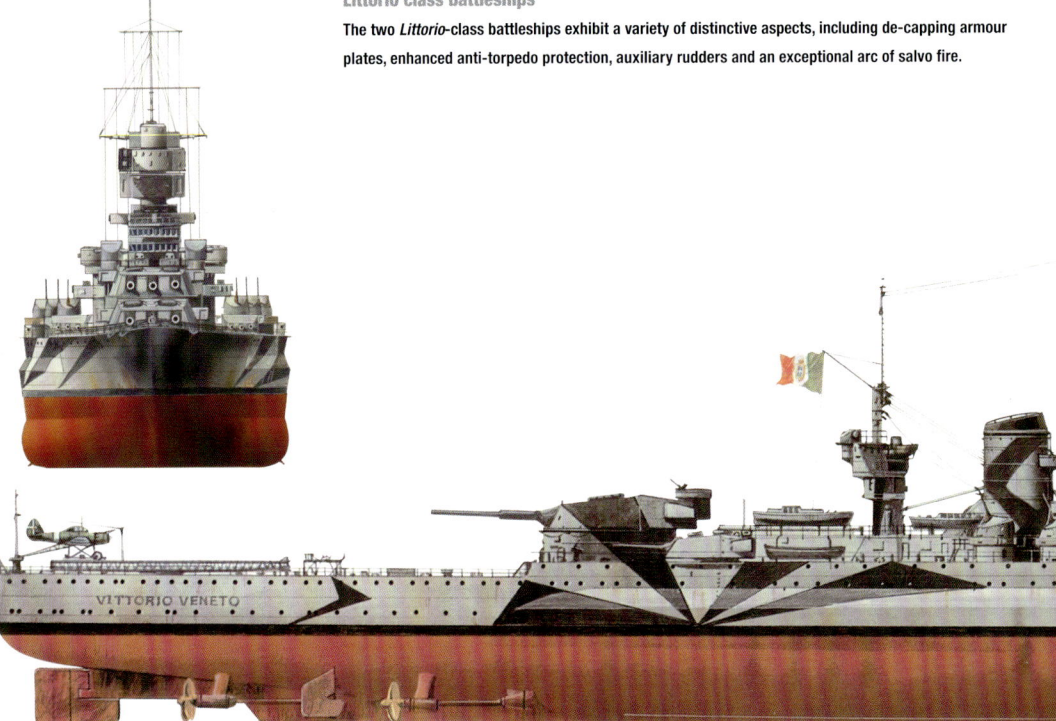

Littorio class battleships
The two *Littorio*-class battleships exhibit a variety of distinctive aspects, including de-capping armour plates, enhanced anti-torpedo protection, auxiliary rudders and an exceptional arc of salvo fire.

WORLD WAR II

muzzle speed of almost 853.5m/s (2800ft/s), with a range of 42.5km (26 mi) at their maximum elevation of 35°. These shells could penetrate 412.5mm (16.5in) armour at 18,288m (20,000yds): comparable to the heavier AP shells fired by USS *North Carolina*'s 406mm (16in) guns.

Unique features

The stern turret was mounted unusually high, enabling it to fire over the 152mm (6in) turret and also giving it an exceptional firing arc of 320°. Uniquely among WWII battleships, the class could pursue a target firing all nine main guns, so long as it was at least 20° off the bow.

The class was fitted with two auxiliary rudders, placed just aft of the outer propellers. Again a unique feature, this was intended to keep the ship manoeuvrable in the event of damage to the main rudder.

Vittorio Veneto was the first Italian battleship to be fitted with radar, Gufo EC4, for surface search, in 1942. The illustration shows the ship in camouflage paint, c.1942.

Vittorio Veneto

Dimensions: Length 237.8m (778ft 9in), Beam 32.9m (107ft 9in), Draught 9.6m (31ft 5in), Displacement 42,040 tonnes (41,377 tons); 46,485 tonnes (45,752 tons) full load

Propulsion: 8 Yarrow boilers, 4 Belluzzo geared turbines, 4 screws, 100,383kW (134,616shp)

Armament: 9 381mm (15in) guns, 12 152mm (6in) guns, 4 120mm (4.7in) guns, 12 90mm (3.5in) AA guns, 20 37mm and 32 20mm AA guns

Armour: Belt 350–60mm (13.8–2.4in), Bulkheads 100–70mm (3.9–2.75in), Deck 205–35mm (8.1–1.4in), Barbettes 350mm (13.8in), Turrets 350–100mm (13.8–3.9in)

Range: 7400km (4000nm) at 16.25 knots (29.6km/h, 18.5mph)

Speed: 31.4 knots (58.2km/h, 36.4mph)

Complement: 1861

Vittorio Veneto at anchor. The Italian battleships initially suffered from a lack of radar, which precluded night action. *Vittorio Veneto* was fitted with surface search radar in mid-1942.

WORLD WAR II

Prince of Wales (1941)

UNITED KINGDOM

The second *King George V* class battleship, *Prince of Wales* had a brief career: it was the first battleship sunk by aircraft on the open sea.

Profile
Prince of Wales as it would have appeared during its Far East deployment in early 1942.

All five of the *King George V* class were laid down in 1937, the *Prince of Wales* at Cammell Laird's yard in Birkenhead in January. Launched on 3 May 1939, it was completed in March 1941. The design was a balanced and handsome one, with two widely spaced funnels, the fore-funnel marginally taller than the aft one, between two tall tripod masts; and the now standard tower structure rising directly behind the forward turrets, with a high-set enclosed navigation bridge. They were the first British battleships to have quadruple main turrets. Based on a 35,560 tonne (35,000 ton) hull, they were to carry 12 356mm (14in) guns in three quadruple turrets. During design, to save weight for use elsewhere, the superfiring 'B' turret was changed to a twin mounting, reducing the broadside to 10 guns.

Main guns
On 31 December 1938, the Washington Naval Treaty ended and was not renewed, although plans had gone ahead in anticipation. In the USA and Japan, 408mm (16in) guns were being assembled for battleships.

Deck plan
The UP projectors, originally fitted on Y turret, were removed mid-1941.

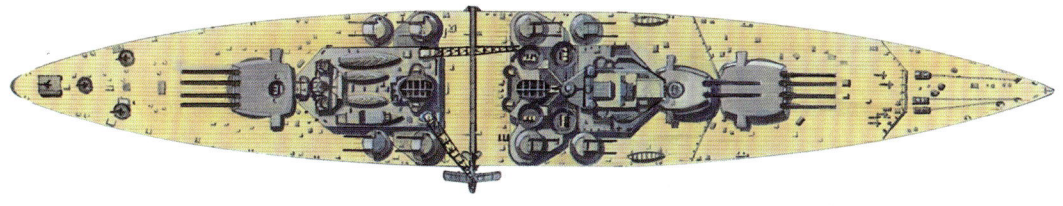

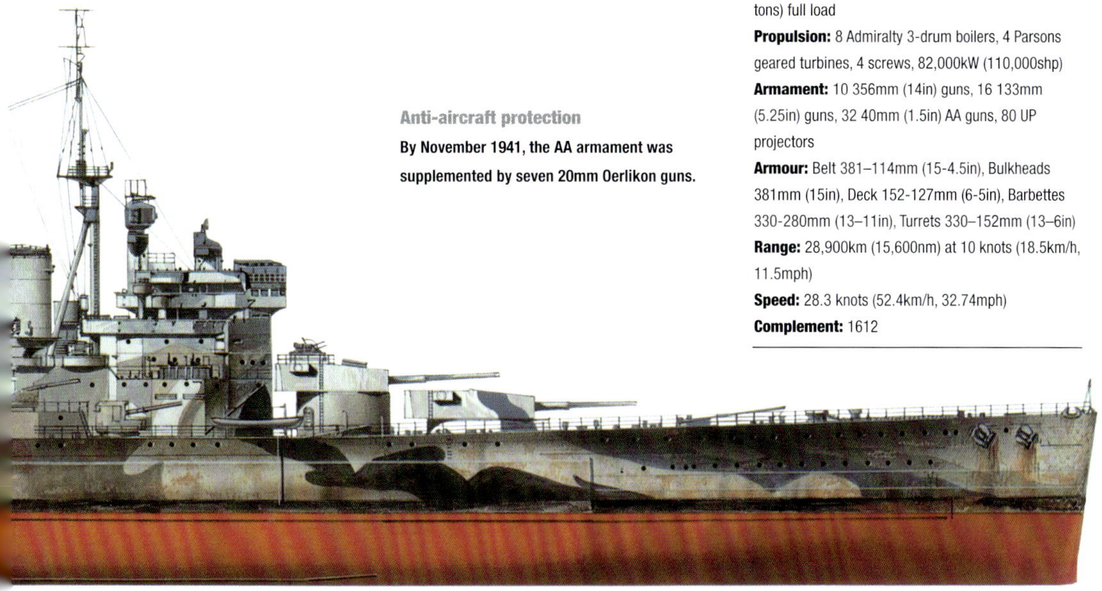

WORLD WAR II

HMS Prince of Wales

Dimensions: Length 227.1m (745ft 1in), Beam 31.4m (103ft 2in), Draught 10.5m (34ft 4in), Displacement 38,610 tonnes (38,000 tons); 45,263 tonnes (44,550 tons) full load
Propulsion: 8 Admiralty 3-drum boilers, 4 Parsons geared turbines, 4 screws, 82,000kW (110,000shp)
Armament: 10 356mm (14in) guns, 16 133mm (5.25in) guns, 32 40mm (1.5in) AA guns, 80 UP projectors
Armour: Belt 381–114mm (15-4.5in), Bulkheads 381mm (15in), Deck 152-127mm (6-5in), Barbettes 330-280mm (13–11in), Turrets 330-152mm (13–6in)
Range: 28,900km (15,600nm) at 10 knots (18.5km/h, 11.5mph)
Speed: 28.3 knots (52.4km/h, 32.74mph)
Complement: 1612

Anti-aircraft protection
By November 1941, the AA armament was supplemented by seven 20mm Oerlikon guns.

Britain had no blueprints for guns of this calibre, but did have a 356mm (14in) quick-firing gun of excellent quality, produced in the early 1930s, and the first of its type to be built in the form of a jacketed and hooped barrel. The turrets posed problems, as it was found that their roller paths flexed with the ship's movements in heavy seas, causing them to jam; the ammunition hoists also had a tendency to stick.

The advent of war prompted numerous alterations during the completion stage, including extra ammunition and fuel capacity, and radar antennae, increasing design displacement to 38,610 tonnes (38,000 tons) and deepening the draught from designed as 10.5m (34ft 4in) to 10.5m (34ft 4in) with a corresponding reduction in freeboard. It was a 'wet ship', requiring wave deflectors on the foredeck.

Secondary armement
Secondary armament was 16 133mm (5.25in) guns, fully automatic and with a range of 20,570m (22,500yds), capable of long-range AA defence as well as knocking out destroyers. Original plans allowed for 32 40mm (1.5in) guns in eight-barrel mountings, and 16 four-barrelled heavy machine

Sea planes
Aircraft were withdrawn from surviving *King George V* class ships in the course of the war.

WORLD WAR II

guns, but the machine guns were not installed on *Prince of Wales*. It was, however, fitted with eight UP smooth-bore multi-barrel AA batteries intended to dangle small aerial mines from parachute wires.

Eight boilers in four compartments provided steam to four independent sets of Parsons geared turbines. The design speed could be forced up to 52.4km/h (28 knots) with an output of 93,212kW (125,000shp).

All or nothing armour

The class carried approximately 12,190 tonnes (12,000 tons) of armour, with a high-level armoured deck resting on top of the side armour, with a maximum thickness of 152mm (6in) above the magazines. Protection was on the 'all or nothing' principle, and armour amounted to 40 per cent of design displacement, based on the realistic awareness that it would have to face 406mm (16in) guns in combat. Careful design of the engine space made it possible to limit the length of the central armoured area to 136m (446ft) on a waterline length of 227m (744ft 9in).

Torpedo bulges were no longer being fitted to Allied battleships, but the torpedo bulkhead was 51mm (2in) and effective torpedo protection space was 4.1m (13ft 4in). The ship had a theoretical immunity zone between 15,727m (17,200yds) and 29,261m (32,000yds) for surface fire. The class marked a jump in speed for Royal Navy battleships, but they were not as fast as some of their rivals.

A double hangar and catapult were provided for Supermarine Walrus amphibious biplanes. These were removed from surviving class ships in the course of the war, made redundant by improved radar.

First combat

Prince of Wales was still engaged on trials when ordered out on its first combat mission with HMS *Hood* against *Bismarck* and *Prinz Eugen* on 24 May 1941. It suffered considerable damage, withdrawing from the battle after the sinking of *Hood*. On 10 December 1941, it was off the east Malayan coast as flagship of Force 'Z' with the battlecruiser *Repulse* and four destroyers, when they were attacked by successive waves of (land-based) Japanese naval aircraft armed with bombs and torpedoes. The ships opened intensive AA fire, increased speed and began zig-zagging, but control of *Prince of Wales*'s port propellers was lost and its manoeuvrability was restricted. After many bomb and torpedo hits, both ships were sunk. The escorting destroyers rescued 1285 men from *Prince of Wales*.

Prince of Wales carried far less AA defence than major ships would do later in the war, although it was probably regarded at the time as well-equipped to deal with air attack, through its High Angle Control System, which was linked to long-range air-search radar. From July 1941 it was carrying radar of Types 279 (air warning), 284 (surface), 282 and 285 (AA control). However, on 10 December 1941, the surface scanning radar was out of action.

All in all, the class did not point the way forward, although the design of Britain's final battleship, HMS *Vanguard*, completed after the war, profited from the lessons learned from the *King George Vs*.

The loss of a brand-new battleship on 10 December 1941 was a severe shock to the Royal Navy and the British public.

WORLD WAR II

Tirpitz (1941)

GERMANY

From 27 May 1941, *Tirpitz* was Nazi Germany's last major battleship, a powerful threat to the enemy, but not to be hazarded against a combined battle group such as that which destroyed *Bismarck*.

Commissioned on 25 February 1941, *Tirpitz* from the start had a raised 'Atlantic' bow, and its profile resemblance to *Bismarck* was very close, although its displacement was slightly greater. *Tirpitz*'s aircraft cranes were fitted on the main deck and it had two double hangars inside the aft superstructure. Both ships had a rotating rangefinder dome mounted above the command tower.

Power and speed
As in *Bismarck*, 12 Wagner superheated high-pressure boilers were fitted, but the drive was from three Brown-Boveri geared turbine sets (acquired from Britain before hostilities), which gave *Tirpitz* slightly greater power and speed despite its greater weight and draught.

Armour protection was essentially the same in both ships. Internally, *Tirpitz* was divided into watertight compartments by 22 transverse bulkheads. The central 'citadel' was closed off fore and aft by bulkheads with a maximum thickness of 220mm (8.6in) and covered by the upper deck (80–50mm, 3.16–1.96in) of Wh (Wotan *hart*) armour topped by 68mm (2.7in) teak planking.

Void spaces of 5.4–3m 17ft 8in–9ft 10in) between the outer hull and inner longitudinal bulkheads of (Ww) Wotan *weich* armour provided anti-torpedo protection.

Like virtually every other World War II warship, *Tirpitz* was fitted with degaussing cables to protect the hull against magnetic mines and torpedoes: its system was known as MSP (magnetic self-protection).

In the winter of 1941–42 alterations were made for the ship's role as guardian of the occupied Norwegian coast and potential convoy raider, including two quadruple torpedo tube mounts and a strengthened AA armament with the original 12 20mm (0.79in) guns being increased to 58.

Radar changes
Rangefinding equipment was installed in the four 380mm (15in) turrets, with both main and secondary guns controllable from the revolving dome

Armament
In the course of the war the armament was supplemented by eight 533mm (21in) torpedo tubes and 58 20mm FlaK30 AA guns.

WORLD WAR II

Deck plan
The deck plan was very similar to that of *Bismarck*.

on the control tower and from the aft control centre. Type SL8 flak gun rangefinders were mounted on each side of the control tower and in midship position aft of the mainmast. The original radar installation was three FuMO23 search sets in 1941, mounted on the forward, foretop and aft rangefinders. By 1944, it had three FuMO26, FuMO Hohentwiel (on the topmast), and FuMO213 WürzburgD fire control mounted on the aft 105mm (4.1in) AA rangefinder.

Like *Bismarck*, *Tirpitz* was a hard ship to destroy. Successive British attacks from 1942 to 1944 failed to neutralize the threat that it presented to Russia-bound convoys. It was only on 12 November 1944 that a force of 32 Lancaster heavy bombers carrying 5.1 tonne (5 ton) 'Tallboy' bombs caused sufficient damage to capsize the ship, with the loss of around 1000 crew.

Tirpitz

Dimensions: Length 251m (823ft 6in), Beam 36m (118ft 1in), Draught 10.61m (34ft 10in, Displacement 43,900 tonnes (43,197 tons); 52,600 tonnes (51,800 tons) full load
Propulsion: 12 Wagner superheated boilers, 3 Brown-Boveri geared turbines, 3 screws, 121,568kW (163,026shp)
Armament: 8 380mm (15in) guns, 12 150mm (5.9in) guns, 16 105mm (4.1in) guns, 16 37mm (1.5in) and 12 20mm (0.79in) AA guns
Armour: Belt 320–80mm (12.6–3.1in), Bulkheads 220–45mm (8.6–1.7in), Deck 120–80mm (4.3–1.7in), Barbettes 340–220mm (13.4–8.6in), Turrets 360–180mm (14.2–7in)
Range: 16,430km (8870nm) at 19 knots (35km/h, 21.9mph)
Speed: 30 knots (57km/h, 35.6mph)
Complement: 2065

Aircraft
Tirpitz carried up to six aircraft, latterly Arado 196A5. The profile above shows the retrieval crane. The profile below shows additional AA guns mounted on turrets Bruno and Cäsar.

WORLD WAR II

Indiana (1942)

UNITED STATES

Indiana (BB-58), second in the four-strong *South Dakota* class, was laid down at Newport News on 20 November 1939 and launched on 21 November 1941. It was completed on 30 April 1942 at a cost of around $77 million.

The superstructure was notably compact, although it included accommodation for a divisional admiral and his staff. The combination of bridge, tower, masts and funnel, varied on each ship and was changed on several occasions. On all the World War II US battleships, the masts became lighter, no longer required for communications or spotting stations. The *South Dakota* class were the first US battleships to have a single funnel from the beginning.

Nine Mark VI 406mm (16in) guns were carried in three triple turrets. Each turret weighed 1,735 tonnes (1,708 tons). Calibre was L/50, the barrels weighed 96.2 tonnes (94.7 tons) and had a range of 38,700m (42,320yds) elevated to 45°. *Indiana*'s secondary armament was 20 127mm (5in) guns in twin turrets, 24 40mm four-barrelled AA guns (a further eight were installed in 1943) and 50 20mm (0.78in) single-mount AA guns (reduced to 40 from 1943).

The immunity zone

Designers were now familiar with the idea of the 'immunity zone' in combat: the area of sea between two ships exchanging long-range fire, with a minimum distance at which the vertical armour plating will defeat an incoming shell and a maximum distance at which horizontal armour (on decks, etc.) will defeat a plunging shell. The zone range was normally reckoned at 18,288–27,432m (20–30,000yds). The *South Dakota's* central armour was designed with these considerations in mind. These were the first US ships to be given inclined internal side armour, reaching from the armoured deck to the inner bottom, 310mm (12.2in) thick, tapering to 25mm (1in). This scheme gave the ships a long, indented

USS Indiana

Dimensions: Length 210m (680ft), Beam 32.9m (107ft 8in), Draught 8.9m (29ft 3in), Displacement 36,476 tonnes (35,900 tons); 45,068 tonnes (44,519 tons) full load
Propulsion: 8 Foster Wheeler boilers, 4 Westinghouse geared turbines, 4 screws, 96,941kW (130,000shp)
Armament: 9 406mm (16in) guns, 20 127mm (5in) guns, 24 40mm and 50 20mm (0.78in) AA guns
Armour: Belt 310–22mm (12.2–0.87in), Bulkheads 279mm (11in), Barbettes 439–287mm (17.3-11.3in), Turrets 457–184mm (18–7.25in), Deck 152–146mm (6–5.75in), Tower 406–184mm (16-7.25in)
Range: 27,750km (15,000nm) at 15 knots (27.75km/h, 17.3mph)
Speed: 27 knots (50km/h, 31.25mph)
Complement: 1793

Floatplanes
Indiana carried two Vought OS2 Kingfisher floatplanes, with a catapult on the fantail.

WORLD WAR II

Indiana at Hampton Roads, Virginia, in September 1942, being prepared and painted for Pacific theatre service.

inward-angled stretch of central hull, just below the flush deck. Another new feature was a splinter protection deck placed 80cm (2ft 7in) below the main armour deck.

Machinery design

Much thought was given to machinery design, as the class was shorter (although of the same beam), than its predecessors – a hull form that did not make for high speed. Eight three-drum Foster Wheeler express boilers were fitted, with two furnaces and double uptakes, working at a pressure of 40.64kg/sq cm (578psi) and a temperature of 454.4°C (850°F) and powering four sets of Westinghouse geared turbines. The engine space was 17m (55ft) shorter than in the preceding *North Carolina*.

Initially all ships carried a rotatory rangefinder on the conning tower, and a Type Sra radar antenna was mounted just abaft. The after mast, mounted on a tripod base, was heightened in 1945. By the end of the war, *Indiana* was fitted with SG surface search radar (aft), and SK-2 air search forward.

War service

Indiana's war service was all with the Pacific Fleet. In September 1946, it was placed on the reserve list, and was the last of the *South Dakota* class to be decommissioned, in September 1947. It was stricken on 1 June 1962 and sold for scrapping on 6 September 1963.

WORLD WAR II

Iowa (1943)

UNITED STATES

Considered by most students as the epitome of the fast battleship, the *Iowa* class was planned from 1938, by which time the constraints of the Washington Treaty had been abandoned. It was going to be the biggest and the best.

Iowa (BB-61) was built at the New York Navy Yard and commissioned on 22 February 1943. Longer than *Indiana* by 64m (70yds) but less than a metre wider (the Panama Canal locks had to be borne in mind), it had a long, narrow forecastle, a distinctive bulbed bow and a broad cruiser-type stern. The long superstructure was packed with guns, aerials and searchlight platforms, with the tower partially enclosing the forward funnel. The masts were mere appendages to the tower and aft funnel (the mainmast was heightened in 1945 and replaced by a light tripod in 1948).

Firepower

The main firepower was nine Mk 7 406mm (16in) guns in three triple turrets. 'Special charge' shells were supplied, to increase muzzle velocity and penetrative effect. At a range of 27,432m (30,000yds), deck armour 190mm (7.6in) thick could be broken through. No warship yet built could withstand that. The secondary armament was 20 Mk 12 127mm (5in) guns in twin mounts. By the time *Iowa* entered service, the vulnerability of capital ships to sustained air attack was very well known, and a massive array of AA guns was fitted. In 1943 it carried 80 40mm (1.6in) Bofors guns and 49 20mm (0.78in) AA Oerlikon cannon, mounted on both sides of the superstructure.

Four reconnaissance floatplanes were carried, launched from two stern-mounted catapults. There was no hangar accommodation and the catapults were removed in 1950 to allow for a helicopter landing pad.

Armour protection was modelled on that developed for the *South Dakota* class, allowing for the increased length. On the protective box around the machinery and magazines,

USS Iowa

Dimensions: Length 270.4m (887ft 2in), Beam 33.5m (108ft 3in), Draught 11.5m (38ft), Displacement 52,834 tonnes (52,000 tons); 58,372 tonnes (57,450 tons) full load
Propulsion: 8 Babcock & Wilcox boilers, 4 GE steam turbines, 4 screws, 158,088kW (212,000shp)
Armament: 9 406mm (16in) guns in 3 turrets, 20 127mm (5in) guns, 60 40mm (1.6in) 4-barrelled AA guns
Armour: Belt 310mm (12.2in), Barbettes 440–287mm (17.3–11.3in), Turret faces 500mm (19.7in), Deck 152mm (6in)
Aircraft: 3, replaced by helicopters, 1949, and UAVs, 1984
Range: 23,960km (12,937nm) at 12 knots (22.2km/h, 13.9mph)
Speed: 33 knots (61.2km/h, 38.25mph)
Complement: 1921

Internal view
Iowa opened up. Masts had become no more than aerial and antenna supports.

WORLD WAR II

Iowa's 406mm (16in) guns in action against shore targets in the Korean War, mid-1952.

the main deck was 38mm (1.5in), the second deck was 121–32mm (4.75–1.25in), splinter deck (above machinery only) 16mm (.62in), third deck 25–16mm (1–.62in), a total of 222mm (8.75in).

Eight high-pressure boilers powered four turbine sets, with the machinery arranged in alternate fire room/turbine room spaces.

Clash of giants?

Radar was installed from the start, although numerous additions were made. In 1943, SK and SRa aerials were fitted on the foremast and in 1945, type SK-2 was added, with type SC-2 on the aft mast. Further additions followed, with necessary alterations to the mast configurations. If *Tirpitz* had sortied into the North Atlantic in autumn 1943, *Iowa* was on station there, but the clash of giants was not to be.

Modernized and with upgraded electronics, it was in action in the Korean War (1949–51) and reclaimed again from reserve in 1984–87 before final decommissioning in 1990. It is now a museum ship in Los Angeles.

115

WORLD WAR II

Missouri (1944)

UNITED STATES

Missouri (BB-63) was last of the *Iowa* class to be actually completed, being commissioned on 11 June 1944 at the New York Navy Yard. In profile and general detail, the ships were very much alike.

Early profile
Missouri's original appearance. Forty years later, it would also carry and fire Harpoon and Tomahawk missiles, along with a Phalanx CIWS anti-missile system.

Detail differences were small but became more noticeable as more antennas were attached to various parts of the superstructure. *Missouri* had an enclosed bridge with straight sides, whereas *Iowa* had a walkway round the traditional open bridge and the conning tower.

Design issues
All shared the slender pointed bow and blunt stern hull, which provided speed and space. One unwelcome feature of the widening hull design was that the ship's 'shoulders' pushed out waves that broke against the sea waves and caused heavy spray to break over the bridge. Another minor but discomforting feature of the hull design was excessive vibration caused by the wake from the inner propellers, which was only partially cured by the adoption of five-bladed screws. Two rudders were fitted, each with a protected area of 31.6sq m (340sq ft). Despite their great length, the class was regarded as easily manoeuvrable. One expert text commented 'at sea these ships handled like destroyers'.

As with the other ships in the class, the main armament was nine Mk 7 406mm (16in) guns in three turrets. Each gun could be aimed and fire independently. The AP shells weighed 1225kg (2700lb) and were fired at a muzzle velocity of 762m/s (2500ft/s); lighter HE shells (862kg/1900lb) were fired at a velocity of 820m/s (2690ft/s). Maximum range was 38.04km (23.64mi). The armour system was intended to provide defence against 406mm (16in) AP shells fired at 768m/s (2520ft/s), the same as *Indiana*.

Propulsion and machinery
Propulsion came from eight Babcock & Wilcox express-type boilers fitted with two furnaces and double uptakes, working at a pressure of 39.72kg/sq cm (565psi) and at a temperature of 454°C (850°F). Superheaters were fitted between each two furnaces. Four General Electric geared turbine sets drove four propellers, 5-bladed inboard and 4-bladed outboard. Arrangement of the machine rooms was the same in all the ships (see *Iowa*) and made for maximum power in minimum space.

There was no longitudinal bulkhead separating the rooms, and a central passageway with a monorail transporter ran the full length of the machinery box,

WORLD WAR II

USS Missouri

Dimensions: Length 270.4m (887ft 3in), Beam 33m (108ft 2in), Draught 11.5m (37ft 9in), Displacement 52,834 tonnes (52,000 tons); 58,372 tonnes (57,540 tons) full load

Propulsion: 8 Babcock & Wilcox boilers, 4 General Electric steam turbines, 4 screws, 158,088kW (212,000shp)

Armament: 9 406mm (16in) guns, 20 127mm (5in) guns, 60 40mm (1.6in) 4-barreled AA guns
Armour: Belt 310mm (12.2in), Barbettes 440–287mm (17.3–11.3in), Turret faces 500mm (19.7in), Deck 152mm (6in)

Range: 28,000km (15,000nm) at 15 knots (28km/h, 17mph)

Speed: 33 knots (61.2km/h, 38mph)

Complement: 1921

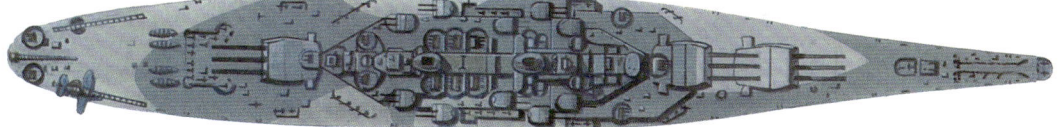

Speed

The *Iowa* class ships were built for high speed, partly to be able to keep up with the Essex-class carriers.

making repair work much easier. The weight of *Missouri*'s machinery was 5072 tonnes (4992 tons) whereas that of *Indiana* was 3637 tonnes (3580 tons). However, *Missouri*'s engines generated 158,088kW (212,000shp), compared to *Indiana's* 96,941kW (130,000shp). Electric power came from eight 1250kW turbogenerators, with two 250kW diesel generators for emergency use. Before World War II, the USN had switched from DC to AC current in shipboard installations.

The engines were reliable: *Missouri* is recorded as steaming continuously for 58 days with no mechanical problems. Bunker capacity was 8765 tonnes (8264 tons) of fuel oil and 190.2 tonnes (187.2 tons) of diesel.

Total crew numbers on this and all other battleships could vary considerably depending, for example, if it was serving as a flagship. However, the tendency was always to rise, mostly because of the greater number of electronic devices and AA guns on

WORLD WAR II

board. The *Iowa* class was originally intended to carry 1921 crew, but in 1945 *Iowa's* number was 2788.

Later service

Missouri is famous as the site of Japan's formal surrender on 31 August 1945. However, between periods in reserve, it was put back in service in 1950–55 and 1984–91, including during the Korean War and Gulf War. In the 1950s a nuclear-capped shell was made specifically for the 406mm (16in) guns. Weighing 862kg (1900lb), its estimated yield was between 15–20 kilotons of TNT. Information on whether these shells were actually carried on the ships remains classified.

1980s modifications

In the 1980s, *Missouri*'s armament was supplemented with missile systems. Launchers for BGM-109 Tomahawk land attack and RGM-84 Harpoon anti-ship rockets were installed, requiring partial rebuilding of the superstructure. The AA guns were removed and replaced with Phalanx CIWS systems. Fire control systems to launch and guide the new weapons were also installed, and radar and communications brought up to date, including an OE82 antenna for satellite communication. The machinery was adapted to burn DFM (diesel fuel marine) rather than NSFO (Navy special fuel oil) fuel, extending periods between boiler maintenance.

Another modern feature was the installation of sewage tanks: previously sewage had been discharged straight into the sea (by all navies). *Missouri* was finally decommissioned on 1 March 1992. It is now a museum ship in Pearl Harbor.

A tug escorts *Missouri* towards the North Island Naval Air Station (CA), 1990.

WORLD WAR II

Guam (1944)

UNITED STATES

The last battlecruisers to be built, the two-ship *Alaska* class was supposed to be an answer to the German Navy's *Scharnhorst* and *Gneisenau*. By the time *Guam* was launched in 1944, the answer was no longer required.

A deck-level gap in the superstructure emphasized the command tower form, with a radar mast attached aft of the platforms. Aircraft hangars were on each side of the tower. Two catapults were fitted in the space between tower and funnel, and three floatplanes were carried (six were originally planned). The emphasis on aircraft is surprising at a time when battleships were losing their planes. Perhaps aerial reconnaissance was regarded as essential to the battlecruiser role. The main navigation bridge was set high, on the seventh level of the command tower. A radar support mast was attached to the aft end of the platforms.

Need for speed

The machinery was designed to produce the high speed demanded, but maximum speed was no greater than that of the new fast battleships. Steam was raised by eight Babcock & Wilcox express boilers. Fitted with superheaters of the same type as the *Iowa*-class ships, they worked at a maximum pressure of 44.6kg/sq cm (634psi). The turbines were of General Electric (GE) cross-compound geared type. From forward, the grouping was fire room, fire room, engine room, repeated. Each engine room had two GE turbogenerators, rated at 1000kW, 450 volt AC, and there were two diesel-powered generators of similar capacity, one at each end of the machinery space.

Nine Mk 8 305mm (12in) guns were carried. The three-gun turret, with each gun elevating independently up to 40°, was preferred for its compactness. Shell weight was 57kg (1140lb) and the muzzle velocity of 762m/s (2500ft/s) gave a maximum range of 35,271m (38,573yds). Magazine capacity was 1500 rounds. The Mk 8 was generally regarded as an excellent gun, capable

USS Guam
Dimensions: Length 264.4m (808ft), Beam 27.8m (91ft 1in), Draught 9.7m (31ft 10in), Displacement 30,257 tonnes (29,779 tons); 34,803 tonnes (34,253 tons) full load
Propulsion: 8 Babcock & Wilcox boilers, 4 General Electric geared turbines, 4 screws, 114,092kW (153,000shp)
Armament: 9 305mm (12in) guns, 12 127mm (5in) AA guns, 56 40mm (1.6in) guns, 34 20mm (0.8in) AA guns
Armour: Belt 229–127mm (9–5in), Turrets 325mm (12.8in), Deck 102mm (4in)
Range: 22,094km (12,000nm) at 15 knots (27.6km/h, 17.26mph)
Speed: 33 knots (61.1km/h, 38mph)
Complement: 1517

of penetrating deck armour 121mm (4.75in) thick at a range of 25,055m (27,400yds). The 127mm (5in) guns could elevate to 85° for AA fire (their number was increased to 68 in 1945); and the 20mm to 90°.

Early profile
USS *Guam* in 1944. *Guam* carried four aircraft amidships, with two laterally-mounted catapults. The most typical planes were Curtiss SC-1 Seahawks.

WORLD WAR II

Armour protection

Large US warships were normally well-protected, especially the *South Dakota* and *Iowa* classes, but the *Alaska* class was armoured only against 305mm (12in) shells. Total thickness of the three armoured decks over the magazines was 169mm (6.65in). Side protection was achieved by a triple hull, with the outer space filled with liquid and the inner one void. While key areas like the steering gear were protected by 256mm (10.6in) plates, the general level of armour was insufficient to withstand heavy shelling.

Radar apparatus consisted of two SG-1 (surface search), SK (air search) and Mk8 GFCS on the main battery gun directors.

Guam, photographed off Trinidad on 13 November 1944 during shakedown training.

Guam entered service in early 1945 and operated with *Alaska* in supporting carrier raids on the Japanese mainland and patrols in the China Sea. Both ships went into reserve after only three years' service. In 1961 they were sold for scrapping.

Glossary

BL: Breech-loading

Barbette: Open-topped armoured enclosure, protecting a gun, with a trunk opening through the armoured floor to the magazine

Beam: Width of a ship

Breastwork: An armoured area raised above deck level, on which guns could be mounted

Cable: Naval unit of measurement, taken variously as 100–120 fathoms (183–219.4m/ 600–720ft)

Calibre: 1. An expression of the inside diameter of the barrel (the bore) and projectile, as in a 305 mm (12-in) gun firing a shell of the same diameter. 2. An expression of the barrel's length as a multiple of the bore: a 305mm (12-in) 40-calibre gun has a barrel length from breech-face to muzzle of 40 times the bore: 12.2m (480in).

Casemate: An armoured area set in the hull, where guns are mounted

Citadel: An area in the centre of a ship, armoured on all sides, enclosing machinery, magazines, etc.

Displacement: A measure of the actual weight of a ship and its contents, excluding cargo

Embrasure: An angled opening in the hull side, enabling a gun to fire forwards or aft

Flash plate: Metal forecastle deck protection on which the anchor chain rests

Howitzer: A short-barreled gun firing a heavy shell at an elevated angle and relatively low muzzle velocity

Hulk: The hull of a vessel with all rigging and equipment removed

Hurricane deck: A raised deck or walkway above the weather deck

Magazine: Secure storage room for gunpowder and explosive shells

ML: Muzzle-loading

MLR: Muzzle-loading, rifled barrel

QF: Quick-firing: guns firing projectiles that combined the shell and the firing charge

Quarterdeck: The open deck aft of the mainmast

Redoubt: An armoured gun emplacement

Reserve: Temporary removal from active service

RB: Rifled bore (gun barrel)

SB: Smoothbore (gun barrel)

Sheer: The fore and aft upward curvature of a ship's hull

Skeg: Sternward extension of the keel as a rudder support

Sponson: A gun-platform extended beyond the side of the hull

Stem: The foremost member of the hull, fixed to the forepart of the keel

Tumblehome: Inward angling of a ship's sides

Turret: An armoured construction containing a gun or guns, able to revolve in a partial or complete circle

Transom: A squared-off stern form

UAV: Unmanned aerial vehicle

Weather deck: The uppermost deck of the ship's hull, open to the sky

Index

References to illustration captions are in **bold**. References to photographs are in *italics*.

A
Adriatic Sea, blockade of 65
Aichi E13A 102
Aichi reconnaissance planes *73*
aircraft
 Britain 35, 49, 55, 67, 71, 81, 84, 109
 France 88, 98
 Germany *78–9*, 86, **91**, 93, 95, 96, 110, **111**
 Japan 73, *73*, **73**, 82, 100, 102
 United States *7*, 50, 60, 74, 77, **112**, 114, 120, **120**
Alaska class 79, 120–1
all or nothing approach to armour 60, 74, 82, 102, 109
Almirante Latorre 56–7, *57*
Amagi 72
Ancona, Italy 41
Andrea Doria 52–3, *53*
Arado Ar 196A **91**, 93
Arado Ar196 floatplanes 86, 96
Archangelsk 67
Arkansas (USS) 38
Armstrong Whitworth shipyard, UK 56
Arsénal de Lorient, France 64
Austria
 Tegetthoff 40–1
 Viribus Unitis 40–1, *41*

B
Bayern 54, 62–3
Beatty, Admiral 35
Bikini Atoll A-bomb tests (1946) 61
Bismarck 80, 81, 84, 94–7, *94*, *96–7*, 98, 109
Blohm & Voss, Hamburg, Germany 94
Blücher 25
Borodino class 17
Brest naval yard, France 65, 88, 98
Bretagne class 64
Britain
 Canada (HMS) 56–7, *57*
 Canopus (HMS) 9, 10–11, *11*
 Courageous (HMS) 70–1
 Dreadnought (HMS) 18–21, *18*, *20*
 Duke of York (HMS) 93
 Hercules (HMS) 32–3, *33*
 Hood (HMS) 6, 80–1, 109
 Indefatigable (HMS) 36
 Indomitable (HMS) 24–5
 Invincible (HMS) 36
 Iron Duke (HMS) 54, 55, 56
 King George V (HMS) 97, 98
 Lion (HMS) 25, 34–7, *37*
 Nelson (HMS) 84
 Prince of Wales (HMS) 79, 106–9, *108–9*
 Queen Elizabeth (HMS) 54–5
 Queen Mary (HMS) 36
 Ramillies (HMS) 48–9
 Repulse (HMS) 79, 109
 Rodney (HMS) 84–5, *85*, 97
 Royal Oak (HMS) 48, *49*
 Royal Sovereign (HMS) 66–9, *67–9*
 Vanguard (HMS) 7, 79, 109
British armed forces
 4th Battle Squadron 21, *33*
 6th Battle Squadron 51
 First Cruiser Squadron 71
 Force 'Z' 109
Burkhardt, Hermann 94, 95

C
Cammell Laird shipyard, Birkenhead, UK 106
Canada (HMS) 56–7, *57*
Canopus (HMS) 9, 10–11, *11*
Cantieri Reuniti dell'Adriatico, Trieste 104
Cape Engaño, Battle of (1944) 73
Capps, Rear Admiral Washington L. 30
Castellammare di Stabia shipyard, Italy 26
Chile, *Almirante Latorre* 56–7, *57*
Colossus class 32
Committee on Naval Design, Britain 19
Courageous class 70
Courageous (HMS) 70–1
Courbet 42–-3, *42*
Cramp's Yard, Philadelphia, US 28, 38

Cuniberti, Vittorio 19, 26
Curtis SOC Seagulls 77

D
D-Day (1944) 51
Dakar 98
Dante Alighieri battleships 40, 41
Danton class 42
Danzig (Gdansk) 23
Derfflinger 6, 36, 44–5, *45*
Deutsche Werke, Kiel 86, 90
Deutsche Werke, Wilhelmshaven 92
Deutschland class 22
Devonport Naval Dockyard, UK 48, 49
D'Eyncourt, Eustace Tennyson 68
Dogger Bank, Battle of (1915) 25, 35–6
Dreadnought (HMS) 18–21, *18, 20*
Duke of York (HMS) 93
Dunkerque 88–9, *89*

F
Fairey III 67
Fairey Flycatchers 81
Fairey Swordfish torpedo bomber 84, 97
Fisher, Admiral Sir John 19, **19**, 24, 34, 48, 66, 70
Fiume (Rijeka) 52
Fore River shipyard, US 60
France
    Courbet 42–-3, *42*

Dunkerque 88–9, *89*
Jauréguiberry 9
Lorraine 65
Provence 64–5, *64*
Richelieu 98–9, *99*
Fusō 6, 58–9

G
Genoa shipyard, Italy 104
Germany
    Bayern 54, 62–3
    Bismarck 80, 81, 84, 94–7, *94, 96*–7, 98, 109
    Blücher 25
    Derfflinger 6, 36, 44–5, *45*
    Deutschland 86–7
    Gneisenau 78–9, 90–1, *90*
    Goeben 34
    Kaiser Friedrich III 9
    Lützow 37, 86–7
    Moltke 34, 35
    Pillau 71
    Prinz Eugen 80, 109
    Scharnhorst 92–3
    Schleswig-Holstein 8–9, 22–3, *22*–3
    Seydlitz 6, 33, 36
    Tirpitz 110–11
Gneisenau 78–9, 90–1, *90*
Goeben 34
Grazhdanin 17
Guam (USS) 120–1, *121*

H
Harvey, Major 37
Heinkel 60c 86
Heinkel HD25 floatplane 82

Heinkel HE114 *78–9*
Heligoland Bight (1917) 71
Hercules (HMS) 32–3, *33*
Hiraga, Yuzuru 82, 100
Hitler, Adolf 94
Hood (HMS) 6, 80–1, 109
Howaldtswerke, Kiel, Germany 62

I
Indefatigable class 34
Indefatigable (HMS) 36
Indiana (USS) 112–13, *113*, 117
Indomitable (HMS) 24–5, 36
Invincible class 24
Invincible (HMS) 36
Iowa class 79, 102, 114–19, 121
Iowa (USS) 9, 101, 114–15, *115*
Iron Duke (HMS) 54, 55, 56
Ise 72–3, *72–3*
Italian-Turkish war (1911) 27
Italy
    Andrea Doria 52–3, *53*
    Littorio 104
    Napoli 27
    Roma 27
    Vittorio Emanuele 26–7, 27
    Vittorio Veneto 104–5, *105*
Iwo Jima, Battle of (1945) 51

J
Japan
    Amagi 72
    Fusō 6, 58–9

INDEX

Ise 72–3, *72–3*
Kongō 46–7, *47*
Mikasa 14–15
Musashi 100
Nagato 82–3
Yamato 6, 100–4, *102–3*
Jauréguiberry 9
Jutland, Battle of (1916) *6, 8–9*, 9, 22, 25, 33, 36, 37, **44**, 79

K
Kaiser class 40
Kaiser Friedrich III 9
Kawasaki, Japan 72
Kearsarge (USS) 12–13, *13*
King George V class 106, **107**
King George V (HMS) 97, 98
Kongō 46–7, *47*
Korean War (1949–51) 115, 118
Krupp cemented armour (KCA) 14
Kure Arsenal, Japan 72, 100

L
La Spezia shipyard, Italy 52
Lancaster heavy bombers 111
lattice masts 28–9
Lexington (USS) 79
Leyte Gulf, Battle of (1944) 73, 83
Lion class 46
Lion (HMS) 25, 34–7, *37*
Littorio 104
Loire 130 seaplanes 88, 98
London Naval Treaty (1930) 7, 100

Lord Nelson class 18
Lorient navy yard, France 42
Lorraine 65
Lützow 37, 86–7
Lyasse, Léon 42

M
Majestic class 10, 14
Malaya 109
Martin FU-1 74
Martin MO-1 74
Martin UO-1 74
Mers-el-Kebir, attack on (1940) 65, 89
Midway, Battle of (1942) 72
Mikasa 14–15
Minotaur class 24, 25
Mississippi (USS) 7, 74–7, 75–7
Missouri (USS) 6, 116–19, *118–19*
Mitsubishi F1M 102
Moltke 34, 35
Montenegro 41
Moon Sound (1917) 17
Musashi 100, 102

N
Nagato 82–3
Napoli 27
Navy List 9
Nelson (HMS) 84
Nevada (USS) 60–1, *61*, 65
New Mexico class 74
New Mexico (USS) 77
New York class 50
New York Navy Yard 114, 116

Newport News, Virginia 12, 50, 112
North Carolina (USS) 105

O
officers' hierarchy 9
Okinawa 51
Oklahoma (USS) 60
Operation Catapult (1940) 65, 89
Operation Torch (1942) 51

P
Pearl Harbor 57, 60, 61
Philippine Sea, Battle of (1944) 83
Pillau 71
Plymouth shipyard, UK 57
Popper, Siegfried 40
Port Arthur, attack on (1904) 16
Port Stanley 11
Portsmouth Navy Yard 18, 66
Prince of Wales (HMS) 79, 106–9, *108–9*
Prinz Eugen 80, 109
Provence 64–5, *64*
Pugliese, Umberto 104

Q
Queen Elizabeth class 44, 62, 66
Queen Elizabeth (HMS) 54–5
Queen Mary (HMS) 36

R
R-class 66
Ramillies (HMS) 48
Regina Elena class 26

République class 16, 26
Repulse (HMS) 79, 109
Revenge class 48
Richelieu 98–9, *99*
Rodney (HMS) 84–5, *85*, 97
Roma 27
Royal Aircraft Factory S.E.5 airplane *7*
Royal Oak (HMS) 48–9, *49*
Royal Sovereign (HMS) 66–9, *67–9*
Russia, *Tsesarevich* 16–17
Russo-Japanese war (1894–95) 17

S
Saneyuki, Akiyama 14
Saratoga (USS) 79
Scapa Flow 21, 38, 49, 63
Scharnhorst 92–3
Schleswig-Holstein 8–9, 22–3, *22–3*
Seydlitz 6, 33, 36
Sopwith 2F Camel 35, 67
Sopwith Pup 35, 50, 67, 71
Sopwith Strutter 35, 71
South Carolina (USS) 19, 28 31, *30–1*
South Dakota class 112, 113, 114, 121
Soviet Union 23
 Archangelsk 67
Spee, Maximilian Graf von 11
Stabilimento Tecnico Triestino 40
Strauss, Joseph 12
Sturdee, Sir Doveton *33*

Supermarine Walrus amphibian planes 84, 109
Surigao Strait, Battle of (1944) 77

T
Tegetthoff 40–1
Texas (USS) 50–1, *51*
Thurston, Sir George 46
Tirpitz 110–11
Togo, Admiral 14
Trieste shipyard 40, 52, 104
Tsesarevich 16–17
Tsushima, Battle of (1905) 14, 15, 26

U
U-29 21, 71
United States
 Arkansas (USS) 38
 Guam (USS) 120–1, *121*
 Indiana (USS) 112–13, *113*, 117
 Iowa (USS) 9, 101, 114–15, *115*
 Kearsarge (USS) 12–13, *13*
 Lexington (USS) 79
 Mississippi (USS) *7*, 74–7, *75–7*
 Missouri (USS) 6, 116–19, *118–19*
 Nevada (USS) 60–1, *61*, 65
 New Mexico (USS) 77
 North Carolina (USS) 105
 Oklahoma (USS) 60
 Saratoga (USS) 79
 South Carolina (USS) 19, 28–31, *30–1*

Texas (USS) 50–1, *51*
Wyoming (USS) 38–9

V
Valsecchi, Vice-Admiral Giuseppe 52
Vanguard (HMS) 7, 79, 109
Vickers shipyard, Barrow-in Furness, UK 14, 46
Virginia class 12
Viribus Unitis 40–1, *41*
Vittorio Emanuele 26–7, *27*
Vittorio Veneto 104–5, *105*
Vought 03U floatplanes 77
Vought OS2 Kingfishers **112**
Vought OS2U Kingfishers 77
Vought UO-1 seaplanes. 74

W
Washington Naval Treaty (1922) 7, 38, 79, 84, 86, 88, 106
White, Sir William 10
Wyoming (USS) 38–9

Y
Yamato 6, 100–4, *102–3*
Yellow Sea, Battle of (1894) 17
Yokosuka D4Y dive bombers **73**
Yokosuka Ro-go Ko-gata floatplane 82

Picture Credits

Artworks:
All Amber Books Ltd.

Photographs:
AirSeaLand.photos: 37 top

Alamy: 6 (Sueddeutsche Zeitung), 8 (CBW), 22 & 23 (Sueddeutsche Zeitung), 64 (Interfoto), 67 (Chronicle), 68/69 (Hilary Morgan), 89 (dpa picture alliance), 108 (GL Archive)

Amber Books: 48, 94, 96/97, 105

Getty Images: 11 (Arkivi), 37 bottom (Popperfoto), 85 (Keystone)

Library of Congess: 13, 18, 51, 53

National Archives and Records Administration: 76, 115

Naval History and Heritage Command: 7, 20, 30/31, 41, 42, 47, 61, 72, 73, 75, 78, 90, 99, 103, 118/119, 121

Public Domain: 27, 33, 45, 57, 113